Mo... ...matics 4

Lawrence Jarrett

A2 Mathematics

Philip Allan Updates, part of the Hodder Education Group, an Hachette Livre UK company, Market Place, Deddington, Oxfordshire OX15 0SE

Orders
Bookpoint Ltd, 130 Milton Park, Abingdon, Oxfordshire OX14 4SB
tel: 01235 827720
fax: 01235 400454
e-mail: uk.orders@bookpoint.co.uk
Lines are open 9.00 a.m.–5.00 p.m., Monday to Saturday, with a 24-hour message answering service. You can also order through the Philip Allan Updates website: www.philipallan.co.uk

© Philip Allan Updates 2007

ISBN 978-1-84489-577-9

First printed 2007
Impression number 5 4 3 2 1
Year 2012 2011 2010 2009 2008 2007

All rights reserved; no part of this publication may be reproduced, stored in a retrieval system, or transmitted, in any form or by any means, electronic, mechanical, photocopying, recording or otherwise without either the prior written permission of Philip Allan Updates or a licence permitting restricted copying in the United Kingdom issued by the Copyright Licensing Agency Ltd, Saffron House, 6–10 Kirby Street, London EC1N 8TS.

This guide has been written specifically to support students preparing for the OCR A2 Mathematics Unit 4724 examination. The content has been neither approved nor endorsed by OCR and remains the sole responsibility of the author.

Typeset by Pantek Arts, Maidstone
Printed by MPG Books, Bodmin

Philip Allan Updates' policy is to use papers that are natural, renewable and recyclable products and made from wood grown in sustainable forests. The logging and manufacturing processes are expected to conform to the environmental regulations of the country of origin.

OCR Unit 4724

Contents

Introduction
About this guide 5
The specification 5
How to answer questions 6
The unit examination 7
How to use this guide 8

Content Guidance
About this section 10
Algebra 11
    Simplification 11
    Division 11
    Partial fractions 12
    Binomial series 14
    General algebra problems 16
Differentiation 17
    Revision 17
    Trigonometric functions 17
    Parametric equations 19
    Implicitly defined equations 21
    General differentiation problems 22
Integration 24
    Revision 24
    Use of partial fractions 25
    Trigonometric functions 27
    Use of trigonometric identities 28
    Expressions of the form $\dfrac{kf'(x)}{f(x)}$ 29
    Integration by parts 31
    Integration by substitution 32
    Choice of correct integration technique 34
    General integration problems 35

Differential equations 36
    General solution 37
    Particular solution 38
    Differential equations in a context 38

Vectors 41
    Basic ideas 41
    Equation of a straight line 42
    Scalar product 44
    General vector problems 45

General Matters

About this section 48

Calculators 49

Formulae booklet 49

Synoptic assessment 51

Questions and Answers

About this section 54

Specimen paper 1 55

Specimen paper 2 62

Specimen paper 3 69

Answers to exercises 74

OCR Unit 4724

Introduction

About this guide

This unit guide is one of a series covering the OCR specification for AS and A-level mathematics. It offers advice for the effective study of **A2 Module 4724: Core Mathematics 4 (C4)**. The guide has four sections:
- **Introduction** — this includes some general points about the specification, about how to revise and about how to tackle the unit examination itself.
- **Content Guidance** — this covers the topic areas in the specification, highlights the key points and provides worked examples. You should attempt to answer each of these worked examples by yourself before checking the given solution and reading the accompanying notes. Each topic is followed by a brief exercise, to give you the chance to consolidate your understanding.
- **General Matters** — this section looks at the appropriate and efficient use of calculators and the formulae booklet and provides information about synoptic assessment.
- **Questions and Answers** — this section contains three specimen papers. Hints and suggestions are provided, together with solutions to the questions. Each solution is followed by an examiner's comment. The answers to the exercises that occur throughout the book are given at the end of the section.

Core Mathematics 4 (C4) is an A2 module. In the examination, some of the questions will be straightforward, routine questions, while others will contain some more challenging aspects, as is appropriate for the final pure mathematics module in A-level mathematics. If you have a thorough understanding of all the topic areas, you will be able to approach the examination with confidence. Of course, hard work is needed; mathematics is best revised by actually doing mathematics, which is why this guide contains plenty of questions for you to attempt.

The specification

The detailed specification for C4 can be found at www.ocr.org.uk. Mathematics cannot be divided into discrete parts, so you will need knowledge learned earlier in your study of mathematics. Indeed the C4 specification states that 'Knowledge of the specification content of Modules C1, C2 and C3 is assumed, and candidates may be required to demonstrate such knowledge in answering questions in Unit C4.' This does not mean that a question in the C4 examination will be based on a topic from an earlier module but it does indicate that, during your solution of a C4 question, you might need to recall a result or technique from C1, C2 or C3.

How to answer questions

Mathematics is a discipline in which accuracy, precision and care are vital. First, read the question carefully — and then re-read it. Adopt a careful and thoughtful approach to setting out your solution; neat and tidy work helps to prevent careless slips. Always double-check that you have written down details from the question accurately in your solution. If you carelessly change the vector $\begin{pmatrix} 2 \\ -4 \\ -7 \end{pmatrix}$ to $\begin{pmatrix} 2 \\ -4 \\ 7 \end{pmatrix}$, or the expression $\dfrac{2x-5}{x^2-5x+7}$ to $\dfrac{2x-5}{x^2-7x+7}$, you will obviously get the wrong answer. Indeed, such an error may prevent you from making any progress with your solution.

Once you think that you have completed the solution, go back to the question to check that you have answered the question that was asked. Make sure that you have answered all the parts of a question. Don't be so ecstatic about having answered parts (i) and (ii) that you overlook the fact that the question also has a part (iii).

Remember that someone is going to look at your solutions. It is in your interests to present your work clearly so that the examiner has an easy job in reading and understanding all aspects of your solutions.

Sketch graphs

A question might require you to sketch a graph. Precise plotting of points is not needed and there is no need to use graph paper. The sketch should be drawn neatly in the answer booklet, showing the basic shape and essential characteristics of the curve.

Part questions

Some questions are divided into parts. If the parts are labelled (a), (b) and (c) then the parts are separate and the answer you obtain in part (a), for example, will be of no relevance to part (b). If the parts are labelled (i), (ii) and (iii), then they may be linked and what you have done in part (i) may have some bearing on what you are to do in part (ii). Look out for words such as 'Hence...' and 'Deduce...'. These signal that the part of the question using such a word definitely follows on from a previous part and that an earlier result will probably be needed.

Exact answers

A question might ask for an exact answer. This means that the method you adopt must be exact throughout. Decimal approximations are not acceptable in such cases, no matter how many decimal places you offer. For example, $\tfrac{1}{2}\pi\sqrt{3}$ is exact but both 0.866π and 2.72 are approximations. The request for an exact answer may be because the examiner wishes to assess a particular method of solution. Therefore, if you resort to decimal values, you are likely to lose several marks.

Significant figures

Unless the question stipulates otherwise, a numerical answer that is not exact should be given correct to 3 significant figures. This does not mean that you do your working using only 3 significant figures. Working must be carried out with greater accuracy (using your calculator as much as possible to retain this accuracy) so that, at the end of the solution, you can judge what is the true 3 significant figure version of the answer.

Proof

A question might ask you to prove or show a result that is quoted in the question. In this case, your solution must be particularly detailed so that you are able to convince the examiner that you know exactly what you are doing.

Sometimes a result is given in the question because it is to be used in a subsequent part of the question. Even if you are unable to prove the given result, you are still entitled — and expected — to use it in answering subsequent parts.

The unit examination

The C4 examination lasts for 1 hour 30 minutes and will probably consist of eight, nine or ten questions. You have to attempt *all* the questions. This is an A2 module, so, while some of the questions will be routine and straightforward, there will be some aspects that are more challenging, designed to test the depth of your mathematical understanding.

The total for the paper is 72 marks. Questions with lower numbers of marks will be towards the beginning of the paper. You should answer the questions in numerical order, leaving a question unfinished only if you are stuck.

The number of marks allocated to a question is a guide to how much is expected. Do not expect to answer a question worth 7 marks in just two lines, while a question worth 2 marks should certainly not take a whole page for the answer. Allowing time for thinking and checking, a total of 72 marks in an examination lasting 90 minutes indicates an approximate rate of 'a mark a minute'.

- Make sure that you arrive at the examination in good time, so that you are calm and fully prepared for the start of the paper.
- Read each question carefully.
- Check that you have transferred details from the question, such as figures and equations, correctly into your solution.
- Tackle each question (including the early, easier ones) in a steady, thoughtful manner — a casual, rushed approach leads to carelessness and to unnecessary loss of marks.

How to use this guide

You will do best if you have a thorough, planned revision programme. Start your revision in good time and set yourself manageable amounts to do each day. This guide contains plenty of questions so that you can practise what you have to do in the examination itself, i.e. answer questions. You do not have to work through this guide in order; a better plan might be to mix topics from algebra, calculus and vectors. Five of the exercises — Exercise 5, Exercise 11, Exercise 20, Exercise 24 and Exercise 28 — consist of more demanding questions. You might like to challenge yourself with these at an appropriate stage.

Once you are confident with all the topics of C4, set aside a period of 90 minutes when you can work without interruption and try one of the specimen papers. The Question and Answer section has further advice on how to approach the papers in the most effective and beneficial way.

Enjoy working steadily and methodically through this guide. I hope you will find the guide helpful and offer my best wishes for the examination.

A2 Mathematics

The content of **A2 Module 4724: Core Mathematics 4 (C4)** consists of five topic areas — algebra, differentiation, integration, differential equations and vectors.

Your algebra skills are developed in this module and rational expressions, i.e. algebraic expressions such as $\dfrac{x^3 + 6x + 7}{2x - 1}$ and $\dfrac{4x + 2}{2x^2 + 2x + 5}$, play a significant part. The algebra topics in C4 are:
- simplification
- division
- partial fractions
- binomial series

A major part of the work in C4 involves calculus. Further approaches to differentiation are considered and several important techniques for integration are included. In disciplines such as biology, economics, chemistry and physics, mathematics is used to describe and analyse situations and, often, differential equations are involved. In C4, differential equations — and one method for solving them — are introduced.

Vector geometry in this module is used to consider the geometry of straight lines in three dimensions. The vector topics in C4 are:
- vector equation of a straight line
- scalar product

The solutions to the worked examples contain explanatory notes, which are preceded by the icon 🛈.

Algebra

Simplification

You should be familiar, from earlier modules, with expanding expressions such as $(2x + 3)(x^2 - 7x + 5)$ and with factorising expressions such as $3x^3 - 10x^2 + 9x - 2$. Rational expressions are algebraic fractions and complicated ones may need to be simplified. This process could involve factorisation and the cancellation of common factors.

> **Worked example**
>
> Simplify $\dfrac{y^2 - 6y + 9}{y^2 - 3y - 4} \div \dfrac{y - 3}{y + 1}$.
>
> **Solution**
>
> 🔲 As in arithmetic, division by a fraction is carried out by multiplying by the reciprocal of the divisor, e.g. $\dfrac{3}{8} \div \dfrac{2}{7} = \dfrac{3}{8} \times \dfrac{7}{2} = \ldots$. To see what cancellation is possible the quadratic expressions must be factorised.
>
> Expression $= \dfrac{y^2 - 6y + 9}{y^2 - 3y - 4} \times \dfrac{y + 1}{y - 3}$
>
> $= \dfrac{(y - 3)^2}{(y - 4)(y + 1)} \times \dfrac{y + 1}{y - 3} = \dfrac{y - 3}{y - 4}$
>
> 🔲 A $(y - 3)$ factor can be cancelled from numerator and denominator, leaving one $(y - 3)$ factor in the numerator; the $(y + 1)$ factors can also be cancelled.

Exercise 1

(1) Simplify $\dfrac{2x^2 + 5x - 3}{4x^2 - 1}$.

(2) Simplify $\dfrac{y^2 - 2y + 1}{2y + 2} \div \dfrac{3y - 3}{5y + 5}$.

(3) Simplify $\dfrac{t^2 - t - 6}{t^2 + 2t} \times \dfrac{t^2 + 5t}{t - 3}$.

Division

You met the process of dividing a cubic polynomial by a linear expression in C2. This is extended in C4 and now it may be a question of dividing a quartic polynomial by a quadratic polynomial. It is important to be aware of the nature of the remainder when carrying out these divisions. When dividing by a linear expression such as $3x + 5$, the remainder will be a number (possibly zero), but when dividing by a quadratic expression, such as $x^2 - 3x + 4$, the remainder could be a linear expression such as $4x + 7$.

> **Worked example**
> Find the quotient and remainder when $x^4 - 3x^3 + x^2 + 4x - 9$ is divided by $x^2 + 2x - 4$.

A2 Mathematics

Solution

📝 A method similar to long division in arithmetic offers the most direct approach.

$$\begin{array}{r}
x^2 - 5x + 15 \\
x^2 + 2x - 4 \overline{\smash{\big)} x^4 - 3x^3 + x^2 + 4x - 9} \\
\underline{x^4 + 2x^3 - 4x^2} \\
-5x^3 + 5x^2 + 4x \\
\underline{-5x^3 - 10x^2 + 20x} \\
15x^2 - 16x - 9 \\
\underline{15x^2 + 30x - 60} \\
-46x + 51
\end{array}$$

Quotient $= x^2 - 5x + 15$; remainder $= -46x + 51$.

📝 Make it clear at the end of the solution which is the quotient and which is the remainder.

Note that, as the result of the division in this example, we can see that the following two identities are equivalent:

$$\frac{x^4 - 3x^3 + x^2 + 4x - 9}{x^2 + 2x - 4} \equiv x^2 - 5x + 15 + \frac{-46x + 51}{x^2 + 2x - 4}$$

$$x^4 - 3x^3 + x^2 + 4x - 9 \equiv (x^2 + 2x - 4)(x^2 - 5x + 15) - 46x + 51$$

Exercise 2

(1) Find the quotient and remainder when $4x^3 + 6x^2 + 9x - 7$ is divided by $x^2 - x + 2$.

(2) Find the quotient and remainder when $x^4 + 7x^3 - 2x - 5$ is divided by $x^2 + 1$.

(3) When the polynomial $p(x)$ is divided by $2x^2 - 3x + 1$, the quotient is $x^2 - 5x - 3$ and the remainder is $12x$. Find $p(x)$.

Partial fractions

Algebraic techniques often involve simplification. One example is the simplification of $\frac{3}{x-2} + \frac{5}{x+1}$, leading to $\frac{3(x+1) + 5(x-2)}{(x-2)(x+1)} = \frac{8x-7}{(x-2)(x+1)}$. This is simplification in the sense that two separate fractions have been added to produce one fractional expression. In some situations, the reverse process is needed and this is called resolving an expression into its partial fractions. It is a technique that is needed, as we shall see later, in integration and in binomial series.

In C4, there are three types of expression which you might be required to resolve into partial fractions:
- a rational expression with two linear factors in the denominator, for example: $\frac{5x-1}{(x-2)(x+3)}$ giving partial fractions of the form $\frac{A}{x-2} + \frac{B}{x+3}$

- a rational expression with three linear factors in the denominator, for example: $\dfrac{8}{x(x+7)(x-3)}$ giving partial fractions of the form $\dfrac{A}{x} + \dfrac{B}{x+7} + \dfrac{C}{x-3}$
- a rational expression with a repeated linear factor in the denominator, for example: $\dfrac{x^2 + 3x + 1}{(x+2)^2 (x-5)}$ giving partial fractions of the form $\dfrac{A}{x+2} + \dfrac{B}{(x+2)^2} + \dfrac{C}{x-5}$

In each case, the values of the constants A, B and C must be found.

Worked example 1

Resolve $\dfrac{15}{(x+2)(2x+3)(x-1)}$ into partial fractions.

Solution

🔢 Different approaches can be taken, but the one used here involves multiplying through by $(x+2)(2x+3)(x-1)$ to produce another identity; then appropriate values for x are substituted in order to find the values of the constants A, B and C.

Let $\dfrac{15}{(x+2)(2x+3)(x-1)} \equiv \dfrac{A}{x+2} + \dfrac{B}{2x+3} + \dfrac{C}{x-1}$

Multiplying through gives $15 \equiv A(2x+3)(x-1) + B(x+2)(x-1) + C(x+2)(2x+3)$
Substituting $x = 1$ gives $15 = A \times 0 + B \times 0 + C \times 3 \times 5$ giving $15 = 15C$ and therefore $C = 1$.
Substituting $x = -\tfrac{3}{2}$ gives $15 = A \times 0 + B \times \tfrac{1}{2} \times \left(-\tfrac{5}{2}\right) + C \times 0$ giving $15 = -\tfrac{5}{4}B$ and therefore $B = -12$.
Substituting $x = -2$ gives $15 = A \times (-1) \times (-3) + B \times 0 + C \times 0$ giving $15 = 3A$ and therefore $A = 5$.

Hence $\dfrac{15}{(x+2)(2x+3)(x-1)} \equiv \dfrac{5}{x+2} - \dfrac{12}{2x+3} + \dfrac{1}{x-1}$

🔢 The complete expression in partial fractions is assembled as the conclusion.

Worked example 2

Express $\dfrac{2x+12}{x^3 + 6x^2 + 9x}$ in partial fractions.

Solution

🔢 First we must factorise the denominator completely. Then we note that this is the type involving a repeated factor in the denominator.

$\dfrac{2x+12}{x^3 + 6x^2 + 9x} = \dfrac{2x+12}{x(x^2 + 6x + 9)} = \dfrac{2x+12}{x(x+3)^2}$

Let $\dfrac{2x+12}{x(x+3)^2} \equiv \dfrac{A}{x} + \dfrac{B}{x+3} + \dfrac{C}{(x+3)^2}$

A2 Mathematics

Multiplying through by $x(x + 3)^2$ gives $2x + 12 \equiv A(x + 3)^2 + Bx(x + 3) + Cx$ *
Substituting $x = -3$ gives $6 = A \times 0 + B \times 0 + C \times (-3)$ giving $C = -2$
Substituting $x = 0$ gives $12 = A \times 9 + B \times 0 + C \times 0$ giving $12 = 9A$ and so $A = \frac{4}{3}$
Comparing coefficients of x^2: $0 = A + B$ giving $B = -\frac{4}{3}$

Hence $\dfrac{2x + 12}{x^3 + 6x^2 + 9x} \equiv \dfrac{\frac{4}{3}}{x} + \dfrac{-\frac{4}{3}}{x + 3} + \dfrac{-2}{(x + 3)^2} \equiv \dfrac{4}{3x} - \dfrac{4}{3(x + 3)} - \dfrac{2}{(x + 3)^2}$

n Note that we multiply through by $x(x + 3)^2$. A common mistake is to multiply through by $x(x + 3)(x + 3)^2$. To find the value of B, we could substitute any other value of x and solve the resulting equation but a neat method is to compare the coefficients of x^2 on both sides of the identity labelled *, noting that there is no term x^2 on the left and anticipating the terms involving x^2 on the right-hand side.

Exercise 3

(1) Resolve $\dfrac{10x + 16}{(2x - 1)(x + 3)}$ into partial fractions.

(2) Express $\dfrac{27}{(x - 2)^2(x + 1)}$ in partial fractions.

(3) Resolve $\dfrac{y^2 + y + 1}{3y^3 + 2y^2 - 8y}$ into partial fractions.

Binomial series

In C2 you met the binomial theorem which gives the expansion of $(a + b)^n$, where n is a positive integer. In C4, this is developed to cases where n is a rational number. The formula is given in the *List of Formulae*:

- $(1 + x)^n = 1 + nx + \dfrac{n(n - 1)}{1 \cdot 2} x^2 + \ldots + \dfrac{n(n - 1)\ldots(n - r + 1)}{1 \cdot 2 \cdot \ldots \cdot r} x^r + \ldots$ $(|x| < 1, n \in \mathbb{R})$

The power n can be a negative integer or a fraction. This is an infinite series and, to have any validity, the terms must become increasingly smaller as the power of x increases; this leads to the restriction $|x| < 1$, i.e. $-1 < x < 1$. It is important to be systematic with this topic — it is easy to make careless errors. In some examples, a preparatory step is necessary so that the expression is of the form $(1 + \ldots)^n$.

Worked example 1

Find the first four terms in the expansion of $(1 - 2x)^{\frac{1}{3}}$ and state the values of x for which the expansion is valid.

Solution

n The first step is to rewrite the expression so that it is of the form $(1 + \ldots)^n$. It is vital in the expansion to remember that the whole $(-2x)$ is to be squared and then cubed.

Rewriting gives: $(1 + (-2x))^{\frac{1}{3}}$

$$(1 + (-2x))^{\frac{1}{3}} = 1 + \frac{1}{3}(-2x) + \frac{\frac{1}{3} \cdot -\frac{2}{3}}{1 \cdot 2}(-2x)^2 + \frac{\frac{1}{3} \cdot -\frac{2}{3} \cdot -\frac{5}{3}}{1 \cdot 2 \cdot 3}(-2x)^3$$

$$= 1 - \frac{2}{3}x - \frac{4}{9}x^2 - \frac{40}{81}x^3$$

The expansion is valid for $-2x$ between -1 and 1, i.e. $-1 < 2x < 1$ and hence $-\frac{1}{2} < x < \frac{1}{2}$.

🛈 We must calculate the coefficients methodically; check the above calculations yourself.

Worked example 2

Expand $\dfrac{1}{(3+x)^2}$ as far as the term in x^2.

Solution

🛈 Great care is needed to prepare this expression for expansion. Be sure you know why the value taken outside the bracket is $\frac{1}{9}$ (and not 9, 3 or $\frac{1}{3}$).

$$\frac{1}{(3+x)^2} = \frac{1}{\left[3\left(1+\frac{1}{3}x\right)\right]^2} = \frac{1}{9\left(1+\frac{1}{3}x\right)^2} = \frac{1}{9}\left(1+\frac{1}{3}x\right)^{-2}$$

$$= \frac{1}{9}\left(1 + (-2)\left(\frac{1}{3}x\right) + \frac{-2 \cdot -3}{1 \cdot 2}\left(\frac{1}{3}x\right)^2 + \ldots\right)$$

$$= \frac{1}{9}\left(1 - \frac{2}{3}x + \frac{1}{3}x^2 + \ldots\right)$$

$$= \frac{1}{9} - \frac{2}{27}x + \frac{1}{27}x^2 + \ldots$$

The expansion is valid when $-1 < \frac{1}{3}x < 1$, i.e. $-3 < x < 3$.

For the expansion of $\dfrac{1}{\sqrt{16+3x}}$, the following are four possible attempts at the first preparatory step:

$$4\left(1+\frac{3}{16}x\right)^{-\frac{1}{2}}, \quad \frac{1}{16}\left(1+\frac{3}{16}x\right)^{-\frac{1}{2}}, \quad \frac{1}{4}\left(1+\frac{3}{16}x\right)^{-\frac{1}{2}}, \quad 16\left(1+\frac{3}{16}x\right)^{-\frac{1}{2}}$$

Which is the correct one?

Exercise 4

(1) Expand $(1+2x)^{-4}$ up to and including the term in x^3.

(2) Find the first three terms in the expansion of $\dfrac{1}{\sqrt{16+3x}}$ in ascending powers of x. State the set of values of x for which the expansion is valid.

(3) Expand $\dfrac{1}{1-\frac{1}{2}x}$ up to and including the term involving x^2. Hence find the coefficient of x^2 in the expansion of $\dfrac{(2+3x)^2}{1-\frac{1}{2}x}$.

General algebra problems

Worked example

It is given that $f(x) = \dfrac{4 - 19x}{4 + 7x - 2x^2}$.

Resolve $f(x)$ into partial fractions and hence find the expansion, in ascending powers of x, up to and including the term in x^2.

Solution

📝 The first step is to factorise the denominator of $f(x)$. Careful rearrangement of the partial fractions is needed before the binomial series expansion can be carried out.

$$f(x) = \dfrac{4 - 19x}{4 + 7x - 2x^2} \equiv \dfrac{4 - 19x}{(4 - x)(1 + 2x)} \equiv \dfrac{A}{4 - x} + \dfrac{B}{1 + 2x}$$

Multiplying through by $(4 - x)(1 + 2x)$ gives $4 - 19x \equiv A(1 + 2x) + B(4 - x)$
Substituting $x = 4$ leads to $A = -8$
Substituting $x = -\tfrac{1}{2}$ leads to $B = 3$

Hence $f(x) = \dfrac{-8}{4 - x} + \dfrac{3}{1 + 2x}$

$= \dfrac{-8}{4(1 - \tfrac{1}{4}x)} + \dfrac{3}{1 + 2x}$

$= -\dfrac{8}{4}\left(1 - \tfrac{1}{4}x\right)^{-1} + 3(1 + 2x)^{-1}$

$= -2\left(1 + \tfrac{1}{4}x + \tfrac{1}{16}x^2\right) + 3(1 - 2x + 4x^2)$

$= 1 - \dfrac{13}{2}x + \dfrac{95}{8}x^2$

📝 Do not forget the importance of rearranging into the $(1 + \ldots)^n$ form before using the binomial series expansion.

Exercise 5

(1) Simplify $\dfrac{t^3 - 9t}{t^3 + 3t^2}$.

(2) Find the first three terms in the series expansion of $\dfrac{\sqrt[3]{1 + 6x}}{(1 - 2x)^2}$.

(3) Resolve $\dfrac{6}{x^3 + 3x^2 - 4}$ into partial fractions.

(4) Expand $\sqrt{4 + x}$ as far as the term in x^2. By substituting $x = \tfrac{1}{4}$, show that $\sqrt{17} \approx \dfrac{2111}{512}$.

(5) (i) Resolve $\dfrac{40}{(2 + x)(1 - 2x)}$ into partial fractions and hence obtain the series expansion for $\dfrac{40}{(2 + x)(1 - 2x)}$ as far as the term in x^2.

 (ii) Given that the coefficient of x^2 in the expansion of $\dfrac{40(1 + kx)}{(2 + x)(1 - 2x)}$ is 5, find the value of the constant k.

Differentiation

Revision

Differentiation is a significant topic in earlier modules and the results and techniques you have already met might be needed in answering C4 questions. The following is a reminder of some of that earlier work.

- Product rule: if $y = uv$, then $\dfrac{dy}{dx} = \dfrac{du}{dx} \cdot v + u \cdot \dfrac{dv}{dx}$

- Quotient rule: if $y = \dfrac{u}{v}$, then $\dfrac{dy}{dx} = \dfrac{v \cdot \dfrac{du}{dx} - u \cdot \dfrac{dv}{dx}}{v^2}$

- Chain rule: if $y = f(g(x))$, then $\dfrac{dy}{dx} = f'(g(x))g'(x)$

- If $y = e^{kx}$, then $\dfrac{dy}{dx} = ke^{kx}$

- If $y = \ln x$, then $\dfrac{dy}{dx} = \dfrac{1}{x}$

Check the accuracy of the following particular differentiation results:

- $y = (4x + 3)^8 \rightarrow \dfrac{dy}{dx} = 32(4x + 3)^7$

- $y = 5e^{-2x} \rightarrow \dfrac{dy}{dx} = -10e^{-2x}$

- $y = x^3 \ln x \rightarrow \dfrac{dy}{dx} = 3x^2 \ln x + x^2$

- $y = (x^3 + 1)^6 \rightarrow \dfrac{dy}{dx} = 18x^2(x^3 + 1)^5$

- $y = \sqrt{8x + 3} \rightarrow \dfrac{dy}{dx} = 4(8x + 3)^{-\frac{1}{2}}$

- $y = \dfrac{5x + 3}{2x - 1} \rightarrow \dfrac{dy}{dx} = -\dfrac{11}{(2x - 1)^2}$

- $y = 6 \ln (x^2 + 2) \rightarrow \dfrac{dy}{dx} = \dfrac{12x}{x^2 + 2}$

Exercise 6

(1) Find the gradient of the curve $y = \dfrac{12}{2x + 5}$ at the point $(-1, 4)$.

(2) Given $f(x) = x^3 e^{2x}$, find $f''(x)$.

(3) Find the equation of the normal to the curve $y = \dfrac{2x + 4}{3x - 1}$ at the point $(1, 3)$.

(4) For the curve $y = 6 \ln (x^2 + 5)$, find the x-coordinates of points on the curve at which the gradient is -2.

Trigonometric functions

New differentiation results in C4 involve trigonometric functions. The basic results are:

- $y = \sin x \rightarrow \dfrac{dy}{dx} = \cos x$

- $y = \cos x \rightarrow \dfrac{dy}{dx} = -\sin x$

- $y = \tan x \to \dfrac{dy}{dx} = \sec^2 x$

Differentiation of trigonometric functions requires x to be measured in radians.

Use of the chain rule gives the following particular results:

- $y = \cos 4x \to \dfrac{dy}{dx} = -4 \sin 4x$
- $y = 3 \sin 5x \to \dfrac{dy}{dx} = 15 \cos 5x$
- $y = \sin(x^3) \to \dfrac{dy}{dx} = 3x^2 \cos(x^3)$
- $y = 8 \tan \tfrac{1}{2}x \to \dfrac{dy}{dx} = 4 \sec^2 \tfrac{1}{2}x$
- $y = \cos^5 x \to \dfrac{dy}{dx} = -5 \cos^4 x \sin x$

Three further results are given in the *List of Formulae*:

- $y = \sec x \to \dfrac{dy}{dx} = \sec x \tan x$
- $y = \operatorname{cosec} x \to \dfrac{dy}{dx} = -\operatorname{cosec} x \cot x$
- $y = \cot x \to \dfrac{dy}{dx} = -\operatorname{cosec}^2 x$

Each of these three further results can be obtained from the earlier results. For example:

$y = \sec x = (\cos x)^{-1} \to \dfrac{dy}{dx} = -1(\cos x)^{-2}(-\sin x)$, using the chain rule

$$= \dfrac{\sin x}{\cos^2 x}$$

$$= \dfrac{1}{\cos x} \times \dfrac{\sin x}{\cos x}$$

$$= \sec x \tan x$$

Worked example 1
Find the gradient of the curve $y = 5 \tan 2x - 3 \sin 4x$ at the point where $x = \tfrac{1}{6}\pi$.

Solution

$\dfrac{dy}{dx} = 10 \sec^2 2x - 12 \cos 4x$

When $x = \tfrac{1}{6}\pi$, $\dfrac{dy}{dx} = 10 \sec^2 \tfrac{1}{3}\pi - 12 \cos \tfrac{2}{3}\pi = \dfrac{10}{\cos^2 \tfrac{1}{3}\pi} - 12 \cos \tfrac{2}{3}\pi$

$$= \dfrac{10}{\left(\tfrac{1}{2}\right)^2} - 12 \times \left(-\tfrac{1}{2}\right) = 46$$

Worked example 2

Find the smallest positive value of x at which the curve $y = e^{3x} \cos 4x$ has a stationary point.

Solution

🔲 Note that differentiation needs the product rule.

$\dfrac{dy}{dx} = 3e^{3x}\cos 4x + e^{3x}(-4\sin 4x)$

For a stationary point, $3e^{3x}\cos 4x - 4e^{3x}\sin 4x = 0 \rightarrow e^{3x}(3\cos 4x - 4\sin 4x) = 0$

e^{3x} can never be zero and so $3\cos 4x - 4\sin 4x = 0$

Dividing through by $\cos 4x$ gives $3 - 4\tan 4x = 0$ and hence $\tan 4x = \tfrac{3}{4}$

∴ $4x = 0.64350...$ and $x = 0.161$

🔲 Remember that your calculator must be in RADIAN mode here. There is no exact answer to provide and so it is appropriate to give the answer correct to 3 significant figures.

Exercise 7

(1) Differentiate each of the following with respect to x:
 (i) $x^2 \tan 5x$
 (ii) $\ln(x - \cos 2x)$
 (iii) $\dfrac{2e^{2x}}{\sin x}$

(2) By writing $\cot x$ as $\dfrac{\cos x}{\sin x}$, use the quotient rule to show that $\dfrac{d}{dx}(\cot x) = -\text{cosec}^2 x$.

(3) Find the equation of the tangent to the curve $y = \sin 2x + 3\cos x$ at the point $\left(\tfrac{1}{6}\pi, 2\sqrt{3}\right)$.

(4) Given that $y = \dfrac{\sin 2x}{1 + \cos 2x}$, show that $\dfrac{dy}{dx} = \dfrac{2}{1 + \cos 2x}$.

Parametric equations

Curves are sometimes defined using parametric equations, the parameter being a third variable such as t or θ. For example, a curve might be defined by:

$x = t^2 + 3t - 4, y = 2t + 1$

Substitution of values for t leads to coordinates:

$t = -3$ $(-4, -5)$
$t = -2$ $(-6, -3)$
$t = -1$ $(-6, -1)$
$t = 0$ $(-4, 1)$
$t = 1$ $(0, 3)$
$t = 2$ $(6, 5)$
$t = 3$ $(14, 7)$

Any real value of t leads to the coordinates of a point. The points may be plotted to give the curve:

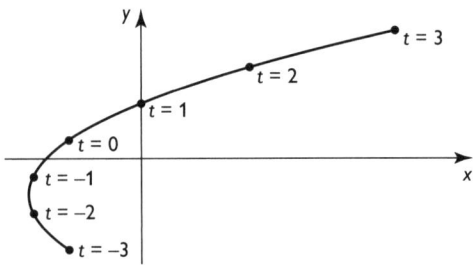

The curve crosses the y-axis where $x = 0$, i.e. $t^2 + 3t - 4 = 0$ giving $t = -4$ and $t = 1$. Substituting these values of t gives the two points $(0, -7)$ and $(0, 3)$. Check that the curve crosses the x-axis at the point $\left(-5\frac{1}{4}, 0\right)$.

For some curves defined parametrically, it is possible, by eliminating the parameter, to produce the cartesian equation of the curve, i.e. an equation which involves only x and y. Show that elimination of t from $x = t^2 + 3t - 4$, $y = 2t + 1$ leads to the cartesian equation $x = \frac{1}{4}y^2 + y - \frac{21}{4}$.

Worked example
Find a cartesian equation of the curve defined by the parametric equations $x = 3 - 2\sin\theta$, $y = 2\cos\theta + 1$.

Solution
◼ The identity $\sin^2\theta + \cos^2\theta \equiv 1$ is the device for eliminating θ.

Rearranging each of the parametric equations gives $\sin\theta = \frac{1}{2}(3 - x)$ and $\cos\theta = \frac{1}{2}(y - 1)$.

Substituting into $\sin^2\theta + \cos^2\theta \equiv 1$ gives $\frac{1}{4}(3 - x)^2 + \frac{1}{4}(y - 1)^2 = 1$, i.e. $(x - 3)^2 + (y - 1)^2 = 4$.

◼ You should recognise this as the equation of a circle, with centre $(3, 1)$ and radius 2.

Exercise 8
(1) A curve is defined by the parametric equations $x = 6t^2 + 1$, $y = \frac{1}{t}$. Find a cartesian equation of the curve.

(2) A curve is defined by the parametric equations $x = t(t + 2)(t + 4)$, $y = t + 5$. Find the coordinates of the points where the curve crosses the axes.

For a curve defined parametrically, it is possible to obtain an expression for the derivative in terms of the parameter. The chain rule means that $\frac{dy}{dx} = \frac{dy}{dt} \times \frac{dt}{dx}$ or, more usefully, $\frac{dy}{dx} = \dfrac{\frac{dy}{dt}}{\frac{dx}{dt}}$

Worked example 1

A curve is defined by $x = t^3 + 2t, y = 4te^{2t}$. Express $\frac{dy}{dx}$ in terms of t.

Solution

Differentiating x gives $\frac{dx}{dt} = 3t^2 + 2$

Differentiating y gives $\frac{dy}{dt} = 4e^{2t} + 8te^{2t}$

Then $\frac{dy}{dx} = \frac{\frac{dy}{dt}}{\frac{dx}{dt}} = \frac{4e^{2t} + 8te^{2t}}{3t^2 + 2}$

🛈 Note that the differentiation of y with respect to t needs the product rule. Always be on the lookout for situations which need the use of the product rule for differentiation.

Worked example 2

A curve is defined by the parametric equations $x = 4\sin 2\theta, y = \theta + \cos\theta$. Find the gradient of the curve at the point for which $\theta = \frac{1}{6}\pi$.

Solution

$\frac{dx}{d\theta} = 8\cos 2\theta$ and $\frac{dy}{d\theta} = 1 - \sin\theta$

The gradient is given by $\frac{dy}{dx} = \frac{1 - \sin\theta}{8\cos 2\theta}$

When $\theta = \frac{1}{6}\pi$, $\frac{dy}{dx} = \frac{1 - \sin\frac{1}{6}\pi}{8\cos\frac{1}{3}\pi} = \frac{1 - \frac{1}{2}}{8 \times \frac{1}{2}} = \frac{1}{8}$

Exercise 9

(1) A curve has parametric equations $x = t^2 + 5t - 2, y = t^2 - t + 4$. Find the equation of the tangent to the curve at the point for which $t = -3$.

(2) A curve has parametric equations $x = 5 - 2\cos\theta, y = 7 + 4\sin\theta$. Show that $\frac{dy}{dx} = 2\cot\theta$.

(3) A curve has parametric equations $x = \frac{1}{4}te^t, y = t^2 - 6t + 1$. Find the value of t for which $\frac{dy}{dx} = 0$ and hence find the coordinates of the stationary point of the curve.

Implicitly defined equations

In C1, you met equations such as $x^2 + y^2 - 4x + 6y - 3 = 0$, which is the equation of a circle. This equation is not of the usual form $y = f(x)$ but is an example of an implicitly defined equation. The equation $x^2 \cos y + y \sin 3x = 16$ is another example; it is impossible to rearrange this equation to the form $y = f(x)$. We need to be able to find expressions for $\frac{dy}{dx}$ and so it is necessary to be certain of how to deal with

differentiating expressions involving both x and y. For example, to differentiate y^3 with respect to x, the chain rule is used:

$$\frac{d}{dx}(y^3) = \frac{d}{dy}(y^3) \cdot \frac{dy}{dx} = 3y^2 \frac{dy}{dx}$$

Check the accuracy of each of the following:

- $\frac{d}{dx}(e^{3y}) = 3e^{3y} \frac{dy}{dx}$
- $\frac{d}{dx}(x^2 \sin y) = 2x \sin y + x^2 \cos y \frac{dy}{dx}$
- $\frac{d}{dx}(y^2 \ln x) = 2y \frac{dy}{dx} \ln x + \frac{y^2}{x}$
- $\frac{d}{dx}(y \tan^2 y) = \frac{dy}{dx} \tan^2 y + y \cdot 2 \tan y \sec^2 y \frac{dy}{dx}$

The product rule is needed for the last three of these results. Note that $\frac{dy}{dx}$ appears every time the differentiation involves a function of y.

Worked example

A curve is defined by $x^3 y + y^3 = 24$. Find the gradient of the curve at the point $(-1, 3)$.

Solution

n The process is simply to differentiate term by term. Note that, once again, the product rule will be involved. Both sides of the equation must be differentiated and, of course, the derivative of the right-hand side is zero.

Differentiating gives $3x^2 y + x^3 \frac{dy}{dx} + 3y^2 \frac{dy}{dx} = 0$

Substituting $x = -1$ and $y = 3$ gives $9 - \frac{dy}{dx} + 27 \frac{dy}{dx} = 0$ and hence $\frac{dy}{dx} = -\frac{9}{26}$

So, the gradient at the point $(-1, 3)$ is $-\frac{9}{26}$.

Exercise 10

(1) A curve has equation $e^{2x} y + y^2 = 6$. Find the equation of the tangent to the curve at the point $(0, -3)$.

(2) A curve is defined by $x^3 + \sin 3y - 8x = 0$. Find an expression in terms of x and y for $\frac{dy}{dx}$.

General differentiation problems

Worked example 1

A curve has equation $y = \sin^2 x - \sin^3 x$ for values of x such that $0 \leqslant x \leqslant \pi$. Find the coordinates of the stationary points.

Solution

⚠ Differentiation of powers of $\sin x$ needs care. For example, rewriting $\sin^3 x$ as $(\sin x)^3$ and using the chain rule shows that the derivative of $\sin^3 x$ is $3(\sin x)^2 \cos x$ or $3\sin^2 x \cos x$.

$\dfrac{dy}{dx} = 2\sin x \cos x - 3\sin^2 x \cos x$

For stationary points, $2\sin x \cos x - 3\sin^2 x \cos x = 0$

Factorising, $\sin x \cos x\,(2 - 3\sin x) = 0$, giving $\sin x = 0$, $\cos x = 0$ or $\sin x = \tfrac{2}{3}$

Hence $x = 0,\ \pi,\ \tfrac{1}{2}\pi,\ 0.730,\ 2.41$

Finding the relevant y-coordinates gives the stationary points as $(0, 0)$, $(\pi, 0)$, $(\tfrac{1}{2}\pi, 0)$, $(0.730, \tfrac{4}{27})$, $(2.41, \tfrac{4}{27})$.

⚠ It would be a mistake to solve the equation $2\sin x \cos x - 3\sin^2 x \cos x = 0$ by dividing through by $\sin x \cos x$ because some of the solutions would be missed. Remember that radians, not degrees, are involved.

Worked example 2

Find the coordinates of each point on the curve $4x^2 + 6xy + y^2 + 10x + 40 = 0$ at which the tangent to the curve is parallel to the y-axis.

Solution

⚠ At a stationary point, the tangent is parallel to the x-axis and $\dfrac{dy}{dx} = 0$. But if a tangent is parallel to the y-axis, $\dfrac{dy}{dx}$ must be infinite, i.e. it is the *denominator* of the expression for $\dfrac{dy}{dx}$ which must be zero.

Differentiating term by term gives $8x + 6y + 6x\dfrac{dy}{dx} + 2y\dfrac{dy}{dx} + 10 = 0$

Rearranging: $(6x + 2y)\dfrac{dy}{dx} = -8x - 6y - 10$

$\dfrac{dy}{dx} = \dfrac{-8x - 6y - 10}{6x + 2y}$

Putting the denominator equal to zero gives $6x + 2y = 0$ or $y = -3x$.
Substituting $y = -3x$ into the equation of the curve to find the relevant points on the curve gives $4x^2 - 18x^2 + 9x^2 + 10x + 40 = 0$, i.e. $-5x^2 + 10x + 40 = 0$.
Solving gives $x = -2$ and 4.
Substituting into $y = -3x$ gives $y = 6$ and $y = -12$.
So, the points where the tangent is parallel to the y-axis are $(-2, 6)$ and $(4, -12)$.

Exercise 11

(1) Find $\frac{dy}{dx}$ for each of the following:

(i) $y = 6 \sin(2x + \frac{1}{3}\pi)$

(ii) $y = \tan^3 x$

(iii) $y = \frac{1}{2} \sec 4x$

(2) A curve is defined by the parametric equations $x = 2e^{3t} + 3t$, $y = 5e^{3t} - 18t$. Find the exact value of t at the point on the curve at which the gradient is 2.

(3) A curve has parametric equations $x = \frac{1}{2} \tan t$, $y = \frac{1}{3} \sec t + 1$. Find a cartesian equation of the curve.

(4) A curve has the equation $4x^2 + y^2 - 16x + 2y - 19 = 0$.

(i) Find the coordinates of the stationary points of the curve.

(ii) Find the coordinates of the points on the curve at which the tangent is parallel to the y-axis.

(5) Given that $f(x) = \frac{\cos 2x}{1 - \sin 2x}$, prove that $f'(x) = \frac{2}{1 - \sin 2x}$ and find the exact value of $f''\left(\frac{1}{12}\pi\right)$.

Integration

Revision

You have met some integration results and applications in C2 and C3 and these might be needed in answering C4 questions. This module develops integration significantly and several important new techniques are included. The following are relevant results from earlier modules:

- $\int x^n \, dx = \frac{1}{n+1} x^{n+1} + c$

- $\int (ax + b)^n \, dx = \frac{1}{a(n+1)} (ax + b)^{n+1} + c$ (where $n \neq -1$)

- $\int e^{kx} \, dx = \frac{1}{k} e^{kx} + c$

- $\int \frac{1}{x} \, dx = \ln|x| + c$

- $\int \frac{1}{ax + b} \, dx = \frac{1}{a} \ln|ax + b| + c$

- Area between a curve and the x-axis $= \int_a^b y \, dx$

- Area between a curve and the y-axis $= \int_c^d x \, dy$

- Volume of region rotated about the x-axis $= \int_a^b \pi y^2 \, dx$

- Volume of region rotated about the y-axis $= \int_c^d \pi x^2 \, dy$

Check the accuracy of the following:

- $\int (9x^2 - 8x + 5)\,dx = 3x^3 - 4x^2 + 5x + c$
- $\int \frac{5}{2x}\,dx = \frac{5}{2}\ln|x| + c$
- $\int 4e^{\frac{1}{2}x}\,dx = 8e^{\frac{1}{2}x} + c$
- $\int_1^9 4\sqrt{x}\,dx = \left[\frac{8}{3}x^{\frac{3}{2}}\right]_1^9 = 72 - \frac{8}{3} = \frac{208}{3}$
- $\int_2^5 \frac{8}{2x-1}\,dx = \left[4\ln|2x-1|\right]_2^5$

$$= 4\ln 9 - 4\ln 3 = 4\ln 3 = \ln 81$$

Exercise 12

(1) Find:
  - **(i)** $\int 6x(x-1)\,dx$
  - **(ii)** $\int \frac{1}{4x+7}\,dx$
  - **(iii)** $\int (2e^{3x} - 8e^{-2x})\,dx$

(2) Evaluate $\int_1^8 (\sqrt[3]{x} + 1)\,dx$.

(3) Find the exact value of $\int_{-2}^{\frac{1}{2}} 4e^{1-2x}\,dx$.

(4) The region enclosed by part of the curve $y = \sqrt{2x+1}$ and the lines $x = 0, x = 5$ and $y = 0$ is rotated completely about the x-axis. Find the volume of the solid produced.

(5) Show that $\int_2^{18} \frac{3}{2x}\,dx = \ln 27$.

Use of partial fractions

We saw earlier in this guide how certain rational expressions can be resolved into partial fractions. For example, on page 13, we saw that:

$$\frac{15}{(x+2)(2x+3)(x-1)} \equiv \frac{5}{x+2} - \frac{12}{2x+3} + \frac{1}{x-1}$$

This alternative form enables integration to be carried out:

$$\int \frac{15}{(x+2)(2x+3)(x-1)}\,dx = \int \left(\frac{5}{x+2} - \frac{12}{2x+3} + \frac{1}{x-1}\right)dx$$

$$= 5\ln|x+2| - 6\ln|2x+3| + \ln|x-1| + c$$

> **Worked example 1**
>
> Resolve $\frac{34-16x}{(2x-3)^2(x+1)}$ into partial fractions and hence find $\int \frac{34-16x}{(2x-3)^2(x+1)}\,dx$.

Solution

n It is important to be sure of the form the partial fractions should take when there is a repeated factor involved.

Let $\dfrac{34 - 16x}{(2x - 3)^2(x + 1)} \equiv \dfrac{A}{(2x - 3)^2} + \dfrac{B}{2x - 3} + \dfrac{C}{x + 1}$

Multiplying through by $(2x - 3)^2 (x + 1)$ gives:
$34 - 16x \equiv A(x + 1) + B(2x - 3)(x + 1) + C(2x - 3)^2$

Substituting $x = \tfrac{3}{2}$ gives $10 = A \times \tfrac{5}{2}$, so $A = 4$.

Substituting $x = -1$ gives $50 = C \times 25$, so $C = 2$.

Comparing coefficients of x^2: $0 = 2B + 4C$, giving $B = -4$.

Hence $\displaystyle\int \dfrac{34 - 16x}{(2x - 3)^2(x + 1)}\, dx = \int \left(\dfrac{4}{(2x - 3)^2} - \dfrac{4}{2x - 3} + \dfrac{2}{x + 1} \right) dx$

$= \displaystyle\int \left(4(2x - 3)^{-2} - \dfrac{4}{2x - 3} + \dfrac{2}{x + 1} \right) dx$

$= -\tfrac{4}{2}(2x - 3)^{-1} - \tfrac{4}{2}\ln|2x - 3| + 2\ln|x + 1| + c$

$= -\dfrac{2}{2x - 3} - 2\ln|2x - 3| + 2\ln|x + 1| + c$

n Note that the integration of $\dfrac{4}{(2x - 3)^2}$ does not involve logarithms. Be sure you understand why this is so.

Worked example 2

Find the exact value of $\displaystyle\int_0^4 \dfrac{3x}{6x^2 + 7x + 2}\, dx$, simplifying the answer.

Solution

n The first step must be to factorise the denominator so that the partial fraction form can be found.

$\displaystyle\int_0^4 \dfrac{3x}{6x^2 + 7x + 2}\, dx = \int_0^4 \dfrac{3x}{(3x + 2)(2x + 1)}\, dx = \int_0^4 \left(\dfrac{6}{3x + 2} - \dfrac{3}{2x + 1} \right) dx$

(applying the usual procedure for resolving into partial fractions)

$= \left[2\ln|3x + 2| - \tfrac{3}{2}\ln|2x + 1| \right]_0^4$

$= \left(2\ln 14 - \tfrac{3}{2}\ln 9 \right) - \left(2\ln 2 - \tfrac{3}{2}\ln 1 \right)$

$= \ln 196 - \ln 27 - \ln 4$

$= \ln \dfrac{196}{27 \times 4}$

$= \ln \dfrac{49}{27}$

n The request for a simplified, exact answer means that some careful work with logarithm properties is needed. There must be no approximate decimal values provided by a calculator.

Exercise 13

(1) Find the exact value of $\int_3^4 \frac{16}{x^2-4}\,dx$.

(2) Find $\int \frac{5x-2}{2x^3-x^2}\,dx$.

(3) Find $\int \frac{5x^2+10x-3}{(x+2)(x+1)(x-1)}\,dx$.

Trigonometric functions

Integration is the reverse process of differentiation so that any differentiation result automatically offers a corresponding integration result. For example, it follows from $\frac{d}{dx}(\sin x) = \cos x$ that $\int \cos x\,dx = \sin x + c$. You need to know the following results:

- $\int \sin kx\,dx = -\frac{1}{k}\cos kx + c$
- $\int \cos kx\,dx = \frac{1}{k}\sin kx + c$
- $\int \sec^2 kx\,dx = \frac{1}{k}\tan kx + c$

Worked example 1

Find $\int (4\cos 2x + 6\sin \tfrac{1}{2}x)\,dx$.

Solution

$$\int (4\cos 2x + 6\sin \tfrac{1}{2}x)\,dx = 4 \cdot \tfrac{1}{2}\sin 2x + 6 \cdot \left(-\frac{1}{\frac{1}{2}}\right)\cos \tfrac{1}{2}x + c$$
$$= 2\sin 2x - 12\cos \tfrac{1}{2}x + c$$

🛈 It is a good idea to differentiate the answer mentally to see that the integrand, $4\cos 2x + 6\sin \tfrac{1}{2}x$, is recovered.

Worked example 2

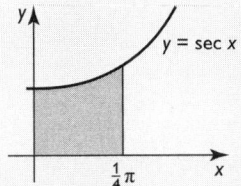

The region shaded in the diagram is rotated completely about the x-axis. Find the volume of the solid produced.

Solution

Volume $= \int_0^{\frac{1}{4}\pi} \pi \sec^2 x\,dx = \left[\pi \tan x\right]_0^{\frac{1}{4}\pi} = \pi \tan \tfrac{1}{4}\pi - \pi \tan 0 = \pi$ cubic units

🛈 Don't forget that, with calculus, radians and not degrees are involved.

Exercise 14

(1) Find $= \int_0^{\frac{1}{6}\pi} 8 \sin 2x \, dx$.

(2) The diagram shows part of the curve $y = 5 \cos \frac{1}{3} x$.

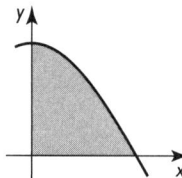

Find the area of the shaded region.

Use of trigonometric identities

An integral such as $\int \sin^2 x \, dx$ cannot be done immediately. (Do not make the mistake of thinking the result is $\frac{1}{3} \sin^3 x + c$ because that differentiates to give $\sin^2 x \cos x$ and not $\sin^2 x$.) However, the use of trigonometric identities can convert some integrals into a form where integration is possible. The identities likely to prove useful in this respect are:

- $\sin^2 x \equiv \frac{1}{2} - \frac{1}{2} \cos 2x$ (which is a rearrangement of $\cos 2x \equiv 1 - 2\sin^2 x$)
- $\cos^2 x \equiv \frac{1}{2} + \frac{1}{2} \cos 2x$
- $\tan^2 x \equiv \sec^2 x - 1$
- $2 \sin x \cos x \equiv \sin 2x$

Remember that these identities can be adjusted to give identities such as:

- $\sin^2 3x \equiv \frac{1}{2} - \frac{1}{2} \cos 6x$
- $\cos^2 \frac{1}{2} x \equiv \frac{1}{2} + \frac{1}{2} \cos x$
- $\tan^2 8x \equiv \sec^2 8x - 1$
- $2 \sin 5x \cos 5x \equiv \sin 10x$

Worked example 1

Find $\int 6 \sin^2 4x \, dx$.

Solution

The adjusted identity will enable us to express $\sin^2 4x$ in terms of $\cos 8x$, producing an expression that can be integrated.

$$\int 6 \sin^2 4x \, dx = \int 6 \left(\frac{1}{2} - \frac{1}{2} \cos 8x \right) dx$$
$$= \int (3 - 3 \cos 8x) \, dx$$
$$= 3x - \frac{3}{8} \sin 8x + c$$

Worked example 2

Evaluate $\int_0^{\frac{1}{4}\pi} \frac{1 + \cos^4 x}{\cos^2 x} dx$.

Solution

🛈 This certainly cannot be integrated immediately, so we must look for a way to convert the integrand to a form that can be integrated.

$$\int_0^{\frac{1}{4}\pi} \frac{1+\cos^4 x}{\cos^2 x} dx = \int_0^{\frac{1}{4}\pi} \left(\frac{1}{\cos^2 x} + \cos^2 x\right) dx = \int_0^{\frac{1}{4}\pi} \left(\sec^2 x + \tfrac{1}{2} + \tfrac{1}{2}\cos 2x\right) dx$$

$$= \left[\tan x + \tfrac{1}{2}x + \tfrac{1}{4}\sin 2x\right]_0^{\frac{1}{4}\pi}$$

$$= \tan \tfrac{1}{4}\pi + \tfrac{1}{8}\pi + \tfrac{1}{4}\sin \tfrac{1}{2}\pi - 0$$

$$= 1 + \tfrac{1}{8}\pi + \tfrac{1}{4}$$

$$= \tfrac{5}{4} + \tfrac{1}{8}\pi$$

🛈 Note that the answer is presented in this exact form involving π.

Exercise 15

(1) Find $\int \cos^2 3x \, dx$.

(2) Find $\int 3\tan^2 \tfrac{1}{2} x \, dx$.

(3) Find the exact value of $\int_0^{\frac{1}{6}\pi} \sin 2x \tan x \, dx$.

Expressions of the form $\frac{kf'(x)}{f(x)}$

Integration is the reverse process of differentiation and each differentiation result corresponds to an integration result. A particular set of integrals exploits this link with differentiation effectively. Consider the following differentiation results, each of which uses the chain rule:

- $y = \ln(x^2 + 5) \rightarrow \frac{dy}{dx} = \frac{2x}{x^2 + 5}$
- $y = \ln(x^3 + 2x + 7) \rightarrow \frac{dy}{dx} = \frac{3x^2 + 2}{x^3 + 2x + 7}$
- $y = \ln(e^{2x} + 1) \rightarrow \frac{dy}{dx} = \frac{2e^{2x}}{e^{2x} + 1}$
- $y = \ln(x^2 + \tan x) \rightarrow \frac{dy}{dx} = \frac{2x + \sec^2 x}{x^2 + \tan x}$

These four results lead to the following:

- $\int \frac{2x}{x^2 + 5} dx = \ln|x^2 + 5| + c$

- $\int \dfrac{3x^2 + 2}{x^3 + 2x + 7}\,dx = \ln|x^3 + 2x + 7| + c$
- $\int \dfrac{2e^{2x}}{e^{2x} + 1}\,dx = \ln|e^{2x} + 1| + c$
- $\int \dfrac{2x + \sec^2 x}{x^2 + \tan x}\,dx = \ln|x^2 + \tan x| + c$

In each case the integrand is of the form $\dfrac{f'(x)}{f(x)}$, i.e. a rational expression in which the numerator is the derivative of the denominator. It follows that $\int \dfrac{f'(x)}{f(x)}\,dx = \ln|f(x)| + c$.

But what about $\int \dfrac{6x + 12}{x^2 + 4x + 7}\,dx$? The numerator is not the derivative of the denominator but they are clearly related because the numerator is 3 times the derivative of the denominator. So $\int \dfrac{6x + 12}{x^2 + 4x + 7}\,dx = 3\int \dfrac{2x + 4}{x^2 + 4x + 7}\,dx = 3\ln|x^2 + 4x + 7| + c$.

The general result is $\int \dfrac{kf'(x)}{f(x)}\,dx = k\ln|f(x)| + c$.

Faced with a rational expression to integrate, it is always worth checking to see if this result can be used.

Worked example

Find $\int \dfrac{12x\cos x - 6x^2 \sin x}{x^2 \cos x + 3}\,dx$.

Solution

The derivative of the denominator, using the product rule of course, is $2x\cos x + x^2(-\sin x)$ or $2x\cos x - x^2 \sin x$ and the integrand is of the form $\dfrac{kf'(x)}{f(x)}$.

$\int \dfrac{12x\cos x - 6x^2 \sin x}{x^2 \cos x + 3}\,dx = 6\int \dfrac{2x\cos x - x^2 \sin x}{x^2 \cos x + 3}\,dx$

$= 6\ln|x^2 \cos x + 3| + c$

This technique also enables us to produce two important results:
- $\int \cot x\,dx = \int \dfrac{\cos x}{\sin x}\,dx = \ln|\sin x| + c$
- $\int \tan x\,dx = \int \dfrac{\sin x}{\cos x}\,dx = (-1)\int \dfrac{-\sin x}{\cos x}\,dx = -\ln|\cos x| + c$

Because $-\ln|\cos x|$ can be written as $\ln|(\cos x)^{-1}|$ or $\ln|\sec x|$, the second of these important results is sometimes given as $\int \tan x\,dx = \ln|\sec x| + c$.

The general results are:
- $\int \cot kx\,dx = \dfrac{1}{k}\ln|\sin kx| + c$
- $\int \tan kx\,dx = -\dfrac{1}{k}\ln|\cos kx| + c = \dfrac{1}{k}\ln|\sec kx| + c$

Rather than trying to memorise these results, it is suggested that you use the general technique, as demonstrated in the following example.

Worked example
Evaluate $\int_0^{\frac{1}{18}\pi} 9\tan 6x\, dx$.

Solution
$$\int_0^{\frac{1}{18}\pi} 9\tan 6x\, dx = \int_0^{\frac{1}{18}\pi} 9 \times \frac{\sin 6x}{\cos 6x}\, dx = -\frac{3}{2}\int_0^{\frac{1}{18}\pi}\frac{-6\sin 6x}{\cos 6x}\, dx$$

$$= \left[-\frac{3}{2}\ln|\cos 6x|\right]_0^{\frac{1}{18}\pi} = -\frac{3}{2}\ln\frac{1}{2} + \frac{3}{2}\ln 1 = \frac{3}{2}\ln 2$$

> Writing down the derivative of $\cos 6x$ immediately in the numerator forces the issue and shows that a factor of $-\frac{3}{2}$ is needed.

Exercise 16
(1) Find $\int \cot 7x\, dx$.

(2) Find $\int \dfrac{5e^{2x} + 10}{e^{2x} + 4x + 3}\, dx$.

(3) Evaluate $\int_0^{\frac{1}{4}\pi} \dfrac{\cos x - \sin x}{\cos x + \sin x + \sqrt{2}}\, dx$.

Integration by parts

To differentiate an expression such as $3x\cos 2x$, the product rule is used. To integrate a similar expression, the process of integration by parts is used. The result, given in the *List of Formulae* booklet, is $\int u\dfrac{dv}{dx}\, dx = uv - \int\dfrac{du}{dx}v\, dx$

Worked example 1
Find $\int 3x\cos 2x\, dx$.

Solution
> It is important to identify u and $\dfrac{dv}{dx}$ carefully and then to find $\dfrac{du}{dx}$ (by differentiation) and v (by integration). Substitution into the formula for integration by parts should then enable the integration to be completed.

So $u = 3x$ and $\dfrac{dv}{dx} = \cos 2x$, giving $\dfrac{du}{dx} = 3$ and $v = \frac{1}{2}\sin 2x$

Substituting into $\int u\dfrac{dv}{dx}\, dx = uv - \int\dfrac{du}{dx}v\, dx$ gives:

$$\int 3x\cos 2x\, dx = 3x \cdot \tfrac{1}{2}\sin 2x - \int 3 \cdot \tfrac{1}{2}\sin 2x\, dx$$

$$= \tfrac{3}{2}x\sin 2x + \tfrac{3}{4}\cos 2x + c$$

Worked example 2

Evaluate $\int_1^e 5x \ln x \, dx$.

Solution

n With $\ln x$ involved, the choice must be to take u as $\ln x$; taking $\frac{dv}{dx}$ as $\ln x$ will not work because we cannot find v immediately.

So, $u = \ln x$ and $\frac{dv}{dx} = 5x$ giving $\frac{du}{dx} = \frac{1}{x}$ and $v = \frac{5}{2}x^2$

Substituting into $\int u \frac{dv}{dx} dx = uv - \int \frac{du}{dx} v \, dx$ gives:

$$\int \ln x \cdot 5x \, dx = \ln x \cdot \tfrac{5}{2} x^2 - \int \tfrac{1}{x} \cdot \tfrac{5}{2} x^2 \, dx$$
$$= \tfrac{5}{2} x^2 \ln x - \int \tfrac{5}{2} x \, dx$$
$$= \tfrac{5}{2} x^2 \ln x - \tfrac{5}{4} x^2 + c$$

So $\int_1^e 5x \ln x \, dx = \left[\tfrac{5}{2} x^2 \ln x - \tfrac{5}{4} x^2 \right]_1^e$

$$= \left(\tfrac{5}{2} e^2 \ln e - \tfrac{5}{4} e^2 \right) - \left(\tfrac{5}{2} \cdot 1 \cdot \ln 1 - \tfrac{5}{4} \right)$$
$$= \tfrac{5}{2} e^2 - \tfrac{5}{4} e^2 + \tfrac{5}{4}$$
$$= \tfrac{5}{4} (e^2 + 1)$$

n Part way through the solution, we are faced with $\int \frac{1}{x} \cdot \frac{5}{2} x^2 \, dx$; this must be simplified to $\int \frac{5}{2} x \, dx$ before the integration is completed.

Exercise 17

(1) Find $\int 8x^3 \ln x \, dx$.

(2) Evaluate $\int_0^\pi 4x \sin \tfrac{1}{2} x \, dx$.

(3) Find $\int 2x e^{-\frac{1}{2}x} \, dx$.

(4) Evaluate $\int_e^{e^2} 4 \ln x \, dx$.

Integration by substitution

The final integration technique in C4 involves a change of variable to convert the integrand into a form that can be integrated. The substitution to be used will always be given in the question, so you do not have to invent a substitution to be used. Indeed, if an examination question does not mention a substitution, then the method of substitution is not the appropriate method. The integral must be completely converted to the new variable — which means the expression, the 'dx' and, if the integration is a definite one, the limits.

Worked example 1

Use the substitution $u = x^3 + 1$ to find $\int \frac{6x^2}{\sqrt{x^3 + 1}} \, dx$.

Solution

◼ $\sqrt{x^3 + 1}$ will become $\sqrt{u}$ but it is not entirely clear how to deal with x^2. So we leave it for the moment in the expectation that things will become clear. The 'dx' must also be dealt with.

Differentiating $u = x^3 + 1$ gives $\dfrac{du}{dx} = 3x^2$ and therefore $dx = \dfrac{du}{3x^2}$.

Substituting: $\displaystyle\int \dfrac{6x^2}{\sqrt{x^3+1}}\,dx = \int \dfrac{6x^2}{\sqrt{u}} \dfrac{du}{3x^2} = \int 2u^{-\frac{1}{2}}\,du = 2 \dfrac{u^{\frac{1}{2}}}{\frac{1}{2}} + c = 4u^{\frac{1}{2}} + c$

$\qquad\qquad\qquad = 4\sqrt{x^3+1} + c$

◼ This is an indefinite integral and so the final step is to return to x for the answer.

Worked example 2

Use the substitution $u = 2x + 1$ to evaluate $\displaystyle\int_1^2 \dfrac{4x+1}{2x+1}\,dx$.

Solution

◼ This is a definite integral so we must change the limits to u values as well.

Differentiating $u = 2x + 1$ gives $\dfrac{du}{dx} = 2$ and therefore $dx = \tfrac{1}{2}\,du$.

Using $u = 2x + 1$, the x limits 1 and 2 become u limits 3 and 5, respectively.

Substituting: $\displaystyle\int_1^2 \dfrac{4x+1}{2x+1}\,dx = \int_3^5 \dfrac{4x+1}{u} \cdot \tfrac{1}{2}\,du$

From $u = 2x + 1$, $2x = u - 1$ and therefore $x = \tfrac{1}{2}(u-1)$

Therefore the integral becomes $\displaystyle\int_3^5 \dfrac{4 \cdot \tfrac{1}{2}(u-1) + 1}{u} \cdot \tfrac{1}{2}\,du = \int_3^5 \dfrac{2u-1}{2u}\,du$

$\qquad\qquad\qquad = \displaystyle\int_3^5 \left(1 - \tfrac{1}{2} \cdot \tfrac{1}{u}\right) du = \left[u - \tfrac{1}{2}\ln|u|\right]_3^5$

$\qquad\qquad\qquad = \left(5 - \tfrac{1}{2}\ln 5\right) - \left(3 - \tfrac{1}{2}\ln 3\right) = 2 - \tfrac{1}{2}\ln \tfrac{5}{3}$

◼ The $4x + 1$ in the numerator has to be converted to an expression in u using the given substitution $u = 2x + 1$. Since the limits have been converted to u values, we must not return to an expression involving x.

Worked example 3

Use the substitution $u = \sin 2x$ to find $\displaystyle\int 4\cos^3 2x\,dx$.

Solution

◼ This example seems strange because $\sin 2x$ does not appear in the integral. But we proceed with the indicated method and expect things to become clear.

Differentiating $u = \sin 2x$ gives $\dfrac{du}{dx} = 2\cos 2x$ and therefore $dx = \dfrac{du}{2\cos 2x}$.

Substituting: $\displaystyle\int 4\cos^3 2x\,dx = \int 4\cos^3 2x \cdot \dfrac{du}{2\cos 2x} = \int 2\cos^2 2x\,du$

Using the identity $\cos^2 2x \equiv 1 - \sin^2 2x$ gives:

integral $= \int 2(1 - \sin^2 2x)\,du = \int 2(1 - u^2)\,du = \int (2 - 2u^2)\,du$

$= 2u - \frac{2}{3}u^3 + c$

$= 2\sin 2x - \frac{2}{3}\sin^3 2x + c$

17 When integrating with respect to u, take care not to use x when you should use u.

Exercise 18

(1) Use the substitution $u = e^{2x} + 3$ to find $\int \dfrac{e^{2x}}{\sqrt{e^{2x} + 3}}\,dx$.

(2) Use the substitution $u = 2x - 1$ to evaluate $\int_{\frac{1}{2}}^{1} 6x(2x - 1)^6\,dx$.

(3) Use the substitution $u = \cos x$ to evaluate $\int_{0}^{\frac{1}{2}\pi} \cos^4 x \sin^3 x\,dx$.

Choice of correct integration technique

Sometimes an examination question testing integration will guide you in the right direction, by stating 'Use integration by parts to...' for example. As already indicated, if the correct technique is substitution, that will be made clear in the question. Often, however, you will have to decide which is the correct technique to adopt. Once that crucial decision is made, the working is often straightforward. The following exercise invites you to indicate the appropriate method to use in each case. For example, for the integral $\int 6xe^{5x}\,dx$, the answer is integration by parts taking $u = 6x$ and $\dfrac{dv}{dx} = e^{5x}$. There is no need to complete each solution and there are no examples requiring integration by substitution.

Exercise 19

Choose the integration technique to be used for each of the following integrals.

(1) $\int \dfrac{8}{(2x + 1)(x - 2)}\,dx$

(2) $\int x^5 \ln x\,dx$

(3) $\int \frac{1}{2}\sin^2 4x\,dx$

(4) $\int 2x \sin 5x\,dx$

(5) $\int \dfrac{x - 1}{x^2 - 2x + 6}\,dx$

(6) $\int \dfrac{x^2 + 1}{x^3 - 4x^2}\,dx$

(7) $\int \ln 3x \, dx$

(8) $\int 4 \sec^2 \tfrac{1}{2} x \, dx$

(9) $\int \dfrac{8}{2x-3} \, dx$

(10) $\int (\cos x + \sin x)^2 \, dx$

(11) $\int \dfrac{1 + \ln x}{4 + x \ln x} \, dx$

(12) $\int x^2 e^{3x} \, dx$

(13) $\int \dfrac{8x}{(x-3)(x+1)(x+2)} \, dx$

(14) $\int 4(3x-1)^6 \, dx$

(15) $\int \dfrac{1 + \sec^3 x}{\sec x} \, dx$

General integration problems

Worked example 1

Find $\int x^2 \sin \tfrac{1}{2} x \, dx$.

Solution

n The obvious technique to try is integration by parts, although it will need two applications before we reach a conclusion.

Choosing $u = x^2$ and $\dfrac{dv}{dx} = \sin \tfrac{1}{2} x$ leads to $\dfrac{du}{dx} = 2x$ and $v = -2 \cos \tfrac{1}{2} x$

Substituting in the formula for integration by parts gives:

$\int x^2 \sin \tfrac{1}{2} x \, dx = x^2(-2 \cos \tfrac{1}{2} x) - \int 2x(-2 \cos \tfrac{1}{2} x) \, dx$

$\qquad = -2x^2 \cos \tfrac{1}{2} x + \int 4x \cos \tfrac{1}{2} x \, dx$

Now, $\int 4x \cos \tfrac{1}{2} x \, dx$ needs integration by parts again.

So choosing $u = 4x$ and $\dfrac{dv}{dx} = \cos \tfrac{1}{2} x$ gives:

$\int 4x \cos \tfrac{1}{2} x \, dx = 4x \cdot 2 \sin \tfrac{1}{2} x - \int 4 \cdot 2 \sin \tfrac{1}{2} x \, dx$

$\qquad = 8x \sin \tfrac{1}{2} x + 16 \cos \tfrac{1}{2} x$

Assembling the complete answer gives:

$\int x^2 \sin \tfrac{1}{2} x \, dx = -2x^2 \cos \tfrac{1}{2} x + 8x \sin \tfrac{1}{2} x + 16 \cos \tfrac{1}{2} x + c$

A2 Mathematics

Worked example 2
Use the substitution $u = x^2$ to evaluate $\int_0^2 x^3 e^{-x^2}\,dx$.

Solution
Differentiating $u = x^2$ gives $\dfrac{du}{dx} = 2x$ and therefore $dx = \dfrac{du}{2x}$.

Using $u = x^2$, the x limits 0 and 2 become u limits 0 and 4 respectively.

Substituting: $\int_0^2 x^3 e^{-x^2}\,dx = \int_0^4 x^3 e^{-u}\,\dfrac{du}{2x} = \int_0^4 \tfrac{1}{2} x^2 e^{-u}\,du = \int_0^4 \tfrac{1}{2} u e^{-u}\,du$

Using integration by parts: integral $= -\tfrac{1}{2} u e^{-u} + \int_0^4 \tfrac{1}{2} e^{-u}\,du = \left[-\tfrac{1}{2} u e^{-u} - \tfrac{1}{2} e^{-u}\right]_0^4$

$= -\tfrac{1}{2} \cdot 4 e^{-4} - \tfrac{1}{2} e^{-4} + \tfrac{1}{2}$

$= \tfrac{1}{2} - \tfrac{5}{2} e^{-4}$

Exercise 20

(1) Find $\int 4x \sec^2 x\,dx$.

(2) Use the substitution $u = 4x - 3$ to evaluate $\int_1^{\frac{5}{4}} \dfrac{16x^2}{4x - 3}\,dx$.

(3) Find $\int 8x^2 e^{2x}\,dx$.

(4) Use the substitution $u = \tan 2x$ to evaluate $\int_0^{\frac{1}{6}\pi} \sec^4 2x\,dx$.

(5) Given that $y = \dfrac{\sin x - \cos x}{\sin x + \cos x}$, find $\dfrac{dy}{dx}$, simplifying your answer. Hence find $\int_0^{\frac{1}{2}\pi} \dfrac{1}{1 + \sin 2x}\,dx$.

(6) Show that $\int_1^2 \dfrac{10x^2 + 9x + 2}{2x^3 + 3x^2 - 1}\,dx = \ln \dfrac{243}{8} - \dfrac{1}{6}$.

(7) Use the substitution $x = 3 \sin \theta$ to evaluate $\int_0^{\frac{3}{2}} \sqrt{9 - x^2}\,dx$.

Differential equations

Differentiation is associated with rates of change. The derivative $\dfrac{dy}{dx}$ is an expression for the rate of change of y with respect to x. Similarly, the derivative $\dfrac{dN}{dt}$ is an expression for the rate of change of N with respect to t.

Because rates of change occur widely in descriptions of processes in science, economics and other disciplines, the study of differential equations is an important topic. A differential equation is simply an equation which includes a derivative. Solving a differential equation means producing an equation in which there is no derivative. Inevitably, integration plays a major part in the solution. In this module, only first-order differential equations in which the variables can be separated are considered.

General solution

A first-order differential equation is one in which the only derivative is a first derivative. The solution involves integration and this leads to the general solution in which there will be one arbitrary constant.

Worked example 1
Find the general solution of the differential equation $\frac{dy}{dx} = \frac{1}{4}x^3 \sec 4y$.

Solution

n We need to separate the variables by legitimate algebraic steps so that the variable y is on one side of the equation and the variable x is on the other.

Separating the variables gives: $\frac{4}{\sec 4y} dy = x^3 dx$

Inserting integration signs: $\int \frac{4}{\sec 4y} dy = \int x^3 dx$ or $\int 4 \cos 4y \, dy = \int x^3 dx$

giving: $\sin 4y = \frac{1}{4}x^4 + c$

n This is the general solution. It could be rewritten $y = \frac{1}{4} \sin^{-1}\left(\frac{1}{4}x^4 + c\right)$ but it is essential to do this only when the question makes the specific request to express the answer as $y = \ldots$.

Worked example 2
Find the general solution of the differential equation $\sqrt{t}\frac{dx}{dt} = 6x$, expressing your answer in the form $x = f(t)$.

Solution

n Letters other than x and y are often used in differential equations and it is important that you are comfortable using other variables in this context. In separating the variables, the '6' could either stay where it is or transfer with x to the left-hand side. Faced with such a choice, always keep the more awkward integral as simple as possible.

Separating the variables gives: $\int \frac{dx}{x} = \int \frac{6}{\sqrt{t}} dt$

giving: $\int \frac{dx}{x} = \int 6t^{-\frac{1}{2}} dt$

Integrating gives $\ln x = 12t^{\frac{1}{2}} + c$

Making x the subject gives $x = e^{12\sqrt{t}+c}$

n This is acceptable as the answer. Note though that it could be written $x = e^{12\sqrt{t}} \cdot e^c$ and, since it is a constant, e^c could be replaced by the constant A, say, giving the answer in the alternative form $x = Ae^{12\sqrt{t}}$.

Exercise 21

(1) Find the general solution of the differential equation $\dfrac{dy}{dx} = \dfrac{4e^{2x}}{y^2}$.

(2) Find the general solution of the differential equation $\dfrac{dy}{dx} = \dfrac{y+2}{\frac{1}{2}x - 3}$, expressing your answer in the form $y = f(x)$.

(3) Find the general solution of the differential equation $\dfrac{d\theta}{dt} = 8te^{2t} \tan\theta$.

Particular solution

The general solution of a differential equation involves an arbitrary constant. Given a piece of extra information, this constant can be determined and the particular solution found.

> **Worked example**
> Find the particular solution of the differential equation $\dfrac{dy}{dx} = \dfrac{x-1}{y+2}$ for which $y = 5$ when $x = 3$.
>
> **Solution**
> First the general solution is found, as in the previous examples. Then the values $x = 3$ and $y = 5$ are substituted to find the value of c.
>
> Separating the variables: $\int (y+2)\,dy = \int (x-1)\,dx$, giving the general solution $\frac{1}{2}y^2 + 2y = \frac{1}{2}x^2 - x + c$.
>
> Substituting $x = 3$, $y = 5$ leads to $\frac{25}{2} + 10 = \frac{9}{2} - 3 + c$ and hence $c = 21$.
>
> The particular solution is $\frac{1}{2}y^2 + 2y = \frac{1}{2}x^2 - x + 21$ or $y^2 + 4y = x^2 - 2x + 42$

Exercise 22

(1) Find the particular solution of the differential equation $\dfrac{dy}{dx} = 4x\sqrt{2y + 3}$ for which $y = 11$ when $x = 3$.

(2) It is given that $(t^2 + 1)\dfrac{dP}{dt} = 2t(P - 1)$ and that $P = 51$ when $t = 3$. Find P in terms of t.

Differential equations in a context

One of the reasons that mathematics is such an important and powerful discipline is its ability to describe — in mathematical terms — situations that arise naturally in many different areas. This means that mathematics plays a significant role in subjects such as biology, chemistry, economics, physics, psychology and sociology. In some of these situations, rates of change are involved and therefore, in each such case, it is a differential equation that provides the description of — or a model for — the 'real-life' process involved.

You could be asked to set up the differential equation that arises from a description involving a rate of change. In general terms, $\frac{dy}{dx}$ represents the rate of change of y with respect to x. Often, the situation involves the rate of change of a variable with respect to time, and then the phrase 'with respect to time' is usually taken for granted. For example, $\frac{dP}{dt}$ represents the rate of change of P and $\frac{dN}{dt}$ represents the rate at which N is changing.

Worked example
A population P is increasing at a rate which, at any instant, is proportional to the population at that instant. Form the differential equation.

Solution
- Remember that, if a variable Y is proportional to a variable X, there exists a constant k (sometimes called the constant of proportionality) such that $Y = kX$.

Rate of change of P is proportional to P, i.e. $\frac{dP}{dt}$ is proportional to P. So there is a constant k such that $\frac{dP}{dt} = kP$.

- You may recall from your study of module C3 that this is the characteristic property of exponential growth.

Exercise 23
(1) The mass, M kg, of a substance is decreasing at a rate which, at a particular instant, is proportional to the mass at that time. Form the differential equation describing this situation.

(2) A liquid is leaking from a container and the rate at which the volume, V cm^3, in the container is decreasing is proportional to the cube root of the volume present. Form the corresponding differential equation.

Worked example
An economist claims that a model for the price, £P, of certain stocks at time t years is the differential equation $\frac{dP}{dt} = \frac{P}{2t+1} + 0.6P\cos 2t$.

(i) Given that $P = 28$ when $t = 0$, find an expression for P in terms of t.
(ii) According to the model, what is the price of the stocks when $t = 5.5$?

Solution
- Before separating the variables, the right-hand side must be factorised. Note that P and $2t + 1$ are necessarily positive and so the modulus signs are not needed with $\ln P$ and $\ln(2t + 1)$.

(i) Differential equation is $\frac{dP}{dt} = P\left(\frac{1}{2t+1} + 0.6\cos 2t\right)$

Separating the variables gives: $\int \frac{1}{P} dP = \int \left(\frac{1}{2t+1} + 0.6\cos 2t\right) dt$

Integrating gives: $\ln P = \frac{1}{2}\ln(2t+1) + 0.3\sin 2t + c$

Substituting $t = 0$, $P = 28$ leads to: $\ln 28 = \frac{1}{2}\ln 1 + 0 + c$, giving $c = \ln 28$

The particular solution is: $\ln P = \frac{1}{2}\ln(2t+1) + 0.3\sin 2t + \ln 28$

Hence $P = e^{\frac{1}{2}\ln(2t+1) + 0.3\sin 2t + \ln 28} = e^{\frac{1}{2}\ln(2t+1)} \cdot e^{0.3\sin 2t} \cdot e^{\ln 28} = 28\sqrt{2t+1}\,e^{0.3\sin 2t}$

(ii) When $t = 5.5$, $P = 28 \times \sqrt{12} \times e^{0.3\sin 11}$

$= 71.8557\ldots$

Model indicates that the price is £71.86.

⚠ Note the precise work needed to express P in terms of t. When substituting $t = 5.5$, remember that your calculator must be in RADIAN mode.

Exercise 24

(1) A population of microorganisms is increasing exponentially, i.e. the rate of increase of the number, N, of microorganisms at time t is proportional to the number present at that time. When $t = 0$, $N = 75$ and when $t = 2$, $N = 192$. Find N in terms of t and hence find the number of microorganisms when $t = 5$.

(2) A boat is leaking oil so that, t hours after the leak starts, the area of the sea affected is A m². A model linking the variables A and t is the differential equation $\frac{dA}{dt} = \frac{1}{5}\sqrt{tA}$. It is given that, when $t = 0$, $A = 1$.
   (i) Find A in terms of t.
   (ii) Find the area of the sea affected by oil after 20 hours.
   (iii) Find the value of t for which the area affected by oil is 250 m².

(3) Find the general solution of the differential equation $\frac{dy}{dx} = \frac{6x}{x^2 y^2 + 3xy^2 + 2y^2}$.

(4) The temperature in a laboratory is 22°C. An object is heated to a temperature of 160°C and then the heat source is removed. The temperature of the object at time t minutes after the removal of the heat source is θ°C. Newton's law of cooling states that the rate at which the object cools is proportional, at any instant, to the difference between the object's temperature and the temperature of the laboratory at that instant.
   (i) Show that $\frac{d\theta}{dt} = k(\theta - 22)$ for some constant k.
   (ii) Find an expression for θ in terms of k and t.
   (iii) 50 minutes after the removal of the heat source, the temperature of the object is 70°C. How much longer will it be before the temperature of the object is 40°C?

Vectors

Basic ideas

Questions on vectors in C4 could involve two dimensions but usually they will involve three dimensions. The origin is O and, with three dimensions involved, there is an x-axis, a y-axis and a z-axis. Each axis is at right angles to each of the other two. Three important vectors are:

- the unit vector in the x-direction indicated by **i** or by $\begin{pmatrix} 1 \\ 0 \\ 0 \end{pmatrix}$

- the unit vector in the y-direction indicated by **j** or by $\begin{pmatrix} 0 \\ 1 \\ 0 \end{pmatrix}$

- the unit vector in the z-direction indicated by **k** or by $\begin{pmatrix} 0 \\ 0 \\ 1 \end{pmatrix}$

Suppose points A and B have coordinates $(2, -1, 5)$ and $(4, 6, -3)$ respectively. The position vector of the point A is $\overrightarrow{OA}$ or **a**, where $\mathbf{a} = 2\mathbf{i} - \mathbf{j} + 5\mathbf{k} = \begin{pmatrix} 2 \\ -1 \\ 5 \end{pmatrix}$. The position vector of the point B is $\overrightarrow{OB}$ or **b**, where $\mathbf{b} = 4\mathbf{i} + 6\mathbf{j} - 3\mathbf{k} = \begin{pmatrix} 4 \\ 6 \\ -3 \end{pmatrix}$.

A simple but vital result in vector work is $\overrightarrow{AB} = \mathbf{b} - \mathbf{a}$, i.e. the vector describing the translation from A to B is found by subtracting **a** from **b**. So here:

$$\overrightarrow{AB} = \begin{pmatrix} 4 \\ 6 \\ -3 \end{pmatrix} - \begin{pmatrix} 2 \\ -1 \\ 5 \end{pmatrix} = \begin{pmatrix} 2 \\ 7 \\ -8 \end{pmatrix}$$

If vector **c** is parallel to vector **d**, then there is a constant p such that $\mathbf{c} = p\mathbf{d}$. Thus the following vectors are parallel to each other: $\begin{pmatrix} 1 \\ -2 \\ 7 \end{pmatrix}, \begin{pmatrix} 3 \\ -6 \\ 21 \end{pmatrix}, \begin{pmatrix} -2 \\ 4 \\ -14 \end{pmatrix}, \begin{pmatrix} \frac{1}{2} \\ -1 \\ \frac{7}{2} \end{pmatrix}$

The magnitude of the vector $\overrightarrow{OA}$ is denoted by $|\overrightarrow{OA}|$. The magnitude of the vector $2\mathbf{i} - \mathbf{j} + 5\mathbf{k}$ is $\sqrt{2^2 + (-1)^2 + 5^2} = \sqrt{30}$.

It is acceptable to use either notation — writing a vector in terms of **i**, **j** and **k** or as a column, but a question could be set using either.

> **Worked example**
> The points P and Q have coordinates $(5, -2, -1)$ and $(11, 0, -4)$ respectively. Find the magnitude of the vector $\overrightarrow{PQ}$.
>
> **Solution**
> *n* The first step uses the useful $\overrightarrow{PQ} = \mathbf{q} - \mathbf{p}$ result. We must be precise in finding the magnitude of any vector — be sure of the process.

$$\overrightarrow{PQ} = \mathbf{q} - \mathbf{p} = \begin{pmatrix} 11 \\ 0 \\ -4 \end{pmatrix} - \begin{pmatrix} 5 \\ -2 \\ -1 \end{pmatrix} = \begin{pmatrix} 6 \\ 2 \\ -3 \end{pmatrix}$$

$$|\overrightarrow{PQ}| = \sqrt{6^2 + 2^2 + (-3)^2} = 7$$

Exercise 25

(1) Points F and G have coordinates (−3, 0, 2) and (1, 3, 7) respectively. Find the magnitude of the vector $\overrightarrow{FG}$.

(2) Points P and Q have coordinates (a, b, −3) and (1, 2, −1) respectively. Given that $\overrightarrow{QP}$ is parallel to the vector $12\mathbf{i} - 5\mathbf{j} + 6\mathbf{k}$, find the values of a and b.

Equation of a straight line

To identify a straight line, two vectors are needed. One is the direction of the line and the other is the position vector of a point that lies on the line. The equation is $\mathbf{r} = \mathbf{a} + \lambda \mathbf{b}$, where:
- **a** is the position vector of a point on the line
- **b** is the vector showing the direction of the line
- λ is the parameter, different values of which give different points on the line

For example, $\mathbf{r} = \begin{pmatrix} 3 \\ 1 \\ 7 \end{pmatrix} + \lambda \begin{pmatrix} 2 \\ -1 \\ 1 \end{pmatrix}$ is the equation of the straight line that is parallel to the vector $\begin{pmatrix} 2 \\ -1 \\ 1 \end{pmatrix}$ and that passes through the point with coordinates (3, 1, 7). The diagram shows that different values of λ give different points on the line:

$\lambda = -1$ (1, 2, 6)
$\lambda = 0$ (3, 1, 7)
$\lambda = 1$ (5, 0, 8)
$\lambda = 2$ (7, −1, 9)
$\lambda = 3$ (9, −2, 10)

Worked example
Find an equation of the straight line through the points (5, 9, −1) and (4, 7, 8).

Solution
11 First find the direction of the line. There is a choice of two points for the position vector of the point that lies on the line; either is correct.

Direction of the line = $\begin{pmatrix} 5 \\ 9 \\ -1 \end{pmatrix} - \begin{pmatrix} 4 \\ 7 \\ 8 \end{pmatrix} = \begin{pmatrix} 1 \\ 2 \\ -9 \end{pmatrix}$

Equation of the line is $\mathbf{r} = \begin{pmatrix} 5 \\ 9 \\ -1 \end{pmatrix} + \lambda \begin{pmatrix} 1 \\ 2 \\ -9 \end{pmatrix}$

n There are different equations that represent the same line. Choosing the point $(4, 7, 8)$ as the point on the line would give the equation $\mathbf{r} = \begin{pmatrix} 4 \\ 7 \\ 8 \end{pmatrix} + \lambda \begin{pmatrix} 1 \\ 2 \\ -9 \end{pmatrix}$ and this is correct too.

There are three possibilities for a pair of straight lines in three dimensions:
- the lines are parallel
- the lines meet at a single point
- the lines are skew (i.e. they are not parallel nor do they meet)

To find out whether two non-parallel lines meet and, if they do, to find their point of intersection, a careful procedure is required:
- Step 1 — form three simultaneous equations involving the parameters of the two lines, say λ and μ.
- Step 2 — solve two of these simultaneous equations to find λ and μ.
- Step 3 — check to see if the values of λ and μ found satisfy the other equation.
  - If this other equation is satisfied, the lines meet.
  - If this other equation is not satisfied, the lines are skew.

Note that, when dealing with two lines in the same question, you must use *different* letters for the two parameters.

Worked example

Show that the lines $\mathbf{r} = \begin{pmatrix} 5 \\ -1 \\ 2 \end{pmatrix} + \lambda \begin{pmatrix} 1 \\ 3 \\ 1 \end{pmatrix}$ and $\mathbf{r} = \begin{pmatrix} 15 \\ 17 \\ 0 \end{pmatrix} + \mu \begin{pmatrix} 2 \\ 3 \\ -1 \end{pmatrix}$ meet and find their point of intersection.

Solution

n First we must carefully write down the three simultaneous equations. One comes from equating the x-coordinates, one from the y-coordinates and one from the z-coordinates.

The equations are $5 + \lambda = 15 + 2\mu$
$$-1 + 3\lambda = 17 + 3\mu$$
$$2 + \lambda = 0 - \mu$$

Solving the first two simultaneous equations gives $\lambda = 2$ and $\mu = -4$.
Now we check in the third equation: left-hand side $= 2 + 2 = 4$
right-hand side $= 0 + 4 = 4$
The third equation is satisfied — so the lines do meet.

Substituting $\lambda = 2$ in the vector equation of the first line gives $\mathbf{r} = \begin{pmatrix} 5 \\ -1 \\ 2 \end{pmatrix} + 2 \begin{pmatrix} 1 \\ 3 \\ 1 \end{pmatrix} = \begin{pmatrix} 7 \\ 5 \\ 4 \end{pmatrix}$

So the point of intersection is $(7, 5, 4)$.

> Putting $\mu = -4$ in the second vector equation will provide a check that the answer is correct.

Exercise 26
(1) Find a vector equation for the line through $(1, -7, 2)$ and $(2, 4, 0)$.
(2) Show that the lines with vector equations $\mathbf{r} = \mathbf{i} + 4\mathbf{j} - 6\mathbf{k} + s(2\mathbf{i} - \mathbf{j} + 3\mathbf{k})$ and $\mathbf{r} = -7\mathbf{i} - \mathbf{j} + 2\mathbf{k} + t(\mathbf{i} + 4\mathbf{j} - 6\mathbf{k})$ are skew.
(3) The line l_1 passes through $(20, -2, 5)$ and $(4, 0, 3)$. The line l_2 has equation $\mathbf{r} = \begin{pmatrix} -6 \\ 11 \\ 0 \end{pmatrix} + \lambda \begin{pmatrix} 1 \\ -5 \\ 1 \end{pmatrix}$.

Show that l_1 and l_2 meet, and find the coordinates of their point of intersection.

Scalar product

The scalar product (sometimes called the dot product) is a way of combining two vectors to produce a scalar quantity, i.e. just a number.
- For two vectors $\mathbf{a}$ and $\mathbf{b}$, the definition of scalar product is $\mathbf{a} \cdot \mathbf{b} = |\mathbf{a}||\mathbf{b}|\cos\theta$, where θ is the angle between the vectors.

Because $\cos 90° = 0$, the scalar product of two perpendicular vectors is zero.

The definition also means that the scalar product of two vectors given in component form can easily be calculated:

$$(2\mathbf{i} + 7\mathbf{j} - 3\mathbf{k}) \cdot (4\mathbf{i} + 5\mathbf{j} + 6\mathbf{k}) = 2 \times 4 + 7 \times 5 + (-3) \times 6 = 25$$

$$\begin{pmatrix} 8 \\ -1 \\ 6 \end{pmatrix} \cdot \begin{pmatrix} -3 \\ -2 \\ 5 \end{pmatrix} = -24 + 2 + 30 = 8$$

An important use of the scalar product is to find the angle between two vectors.

Worked example
Find the angle θ between the straight lines $\mathbf{r} = 3\mathbf{i} - 7\mathbf{j} + \mathbf{k} + \lambda(\mathbf{i} + 2\mathbf{j} - 3\mathbf{k})$ and $\mathbf{r} = \mathbf{i} + \mathbf{j} + \mu(-\mathbf{i} + 5\mathbf{j} - 2\mathbf{k})$.

Solution
> The angle between the lines depends solely on their directions; it does not matter whether the lines intersect or not. So here, we need to find the angle between the vectors $\mathbf{i} + 2\mathbf{j} - 3\mathbf{k}$ and $-\mathbf{i} + 5\mathbf{j} - 2\mathbf{k}$.

Using the definition, scalar product $= \sqrt{1^2 + 2^2 + (-3)^2}\sqrt{(-1)^2 + 5^2 + (-2)^2}\cos\theta$
$= \sqrt{14}\sqrt{30}\cos\theta$

But also $(\mathbf{i} + 2\mathbf{j} - 3\mathbf{k}) \cdot (-\mathbf{i} + 5\mathbf{j} - 2\mathbf{k}) = -1 + 10 + 6 = 15$

Hence $\sqrt{14}\sqrt{30}\cos\theta = 15$, giving $\cos\theta = \dfrac{15}{\sqrt{14}\sqrt{30}}$ and $\theta = 43.0°$

Exercise 27

(1) Given that the vectors $\begin{pmatrix} a \\ -1 \\ 4 \end{pmatrix}$ and $\begin{pmatrix} 3 \\ 2a \\ -2 \end{pmatrix}$ are perpendicular, find the value of a.

(2) Find the acute angle between the vectors $3\mathbf{i} - 5\mathbf{j} + 2\mathbf{k}$ and $-2\mathbf{i} + \mathbf{j} + 2\mathbf{k}$.

(3) Points P and Q have coordinates $(5, 1, 4)$ and $(-2, 0, 3)$ respectively. Find the acute angle between PQ and the line with equation $\mathbf{r} = \mathbf{i} + \mathbf{k} + \lambda(2\mathbf{i} - 3\mathbf{j})$.

General vector problems

Worked example

Find the size of the angle A in the triangle ABC shown.

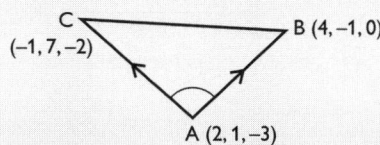

Solution

n The angle could be acute or obtuse and so it is important to use the scalar product in a careful and consistent manner — by considering both relevant vectors with A as the starting point, i.e. to consider the scalar product $\vec{AB} \cdot \vec{AC}$.

$\vec{AB} = \begin{pmatrix} 4 \\ -1 \\ 0 \end{pmatrix} - \begin{pmatrix} 2 \\ 1 \\ -3 \end{pmatrix} = \begin{pmatrix} 2 \\ -2 \\ 3 \end{pmatrix}$

$\vec{AC} = \begin{pmatrix} -1 \\ 7 \\ -2 \end{pmatrix} - \begin{pmatrix} 2 \\ 1 \\ -3 \end{pmatrix} = \begin{pmatrix} -3 \\ 6 \\ 1 \end{pmatrix}$

$\sqrt{4+4+9}\,\sqrt{9+36+1}\,\cos A = -6 - 12 + 3$, giving $\cos A = \dfrac{-15}{\sqrt{17}\,\sqrt{46}}$ and so $A = 122.4°$.

n Note that the negative value of cos A indicates that the angle A is obtuse.

Exercise 28

(1) Points D, E and F have coordinates $(11, -2, 7)$, $(4, 1, 9)$ and $(-2, -3, 1)$ respectively. Find the coordinates of the point G such that $DEFG$ is a parallelogram.

(2) Calculate the acute angle between the y-axis and the line with equation $\mathbf{r} = \lambda(4\mathbf{i} - 2\mathbf{j} + 7\mathbf{k})$.

(3) It is given that the lines $\mathbf{r} = \begin{pmatrix} 2 \\ -11 \\ 1 \end{pmatrix} + \lambda\begin{pmatrix} -1 \\ m \\ 5 \end{pmatrix}$ and $\mathbf{r} = \begin{pmatrix} 13 \\ 3 \\ p \end{pmatrix} + \mu\begin{pmatrix} 8 \\ -3 \\ m \end{pmatrix}$ intersect and are perpendicular. Find the values of the constants m and p and find the coordinates of the point of intersection.

(4) Show that the line joining the points $(1, -1, 7)$ and $(-15, -1, 19)$ meets the line joining the points $(10, -2, 7)$ and $(13, -5, 25)$, and find the coordinates of the point of intersection. Find also the acute angle between the two lines.

A2 Mathematics

This section looks at three important general aspects of preparing to take the
A2 Module 4724: Core Mathematics 4 (C4) examination:
- calculators
- the formulae booklet
- the relevance of previous mathematical knowledge

Calculators

In the C4 examination, you are allowed to use a scientific calculator, a graphical calculator, or both. However, you must *not* use a calculator that possesses computer algebra functions. You are expected to use your calculator accurately and efficiently; however, you must also understand when it is not appropriate to use your calculator.

Much of the work in this module involves exact values and so calculators will be of limited use. Note, though, that they can be useful as a means of checking an exact answer. For example, suppose that, given the curve with equation $y = 5\cos x - 2\sin 2x + \tan 5x$, you are required to find the exact value of y when $x = \frac{1}{6}\pi$. This leads to:

$$5\cos\tfrac{1}{6}\pi - 2\sin\tfrac{1}{3}\pi + \tan\tfrac{5}{6}\pi$$
$$= 5 \cdot \tfrac{1}{2}\sqrt{3} - 2 \cdot \tfrac{1}{2}\sqrt{3} - \tfrac{1}{\sqrt{3}}$$
$$= \tfrac{5}{2}\sqrt{3} - \sqrt{3} - \tfrac{1}{3}\sqrt{3}$$
$$= \tfrac{7}{6}\sqrt{3}$$

This is the required exact value and your calculator will probably not be able to provide this. But you can use your calculator, set to RADIAN mode, to confirm that $5\cos\tfrac{1}{6}\pi - 2\sin\tfrac{1}{3}\pi + \tan\tfrac{5}{6}\pi$ and $\tfrac{7}{6}\sqrt{3}$ do give the same decimal approximation of 2.020 725 942.

It is not essential to use a graphical calculator, although careful and informed use of one should enhance your understanding of mathematics. One of the topics in C4 is parametric equations and a graphical calculator can display a curve defined by parametric equations. As with all graphs displayed on a calculator, the key features of the curve are only shown effectively if the scales are set appropriately.

Exercise 29

(1) A curve is defined by $y = e^{5x}\cos 2x$ for $0 \leq x \leq \tfrac{1}{2}\pi$. Find the coordinates of the stationary point, giving each coordinate correct to 5 decimal places.

(2) A curve is defined by the parametric equations $x = 2 + 3\sin\theta$, $y = 1 + 4\cos\theta$. Use a graphical calculator to provide a sketch of the curve.

(3) Find the exact value of $\int_{0}^{\frac{1}{6}\pi}(4\cos 2x + \sec^2 x)\,dx$.

Formulae booklet

The *List of Formulae* booklet (List MF1) is an important resource in the examination. It is essential that you are familiar with this booklet and how it can be of use during the C4 examination. The same booklet is used for all the OCR mathematics examinations, so it contains formulae and statistical tables for all units. It is a good idea to use it while preparing for the examination, so that you can easily find any formula you need. You can check the website **www.ocr.org.uk** for a copy of the booklet.

The booklet contains many formulae that you will have met while studying the modules C1, C2, C3 and C4. The formulae in the booklet of most relevance to C4 are:

- $(1 + x)^n = 1 + nx + \dfrac{n(n-1)}{1 \cdot 2} x^2 + \ldots + \dfrac{n(n-1)\ldots(n-r+1)}{1 \cdot 2 \ldots r} x^r + \ldots \quad (|x| < 1, n \in \mathbb{R})$
- $f(x) = \tan kx \rightarrow f'(x) = k \sec^2 kx$
- $f(x) = \sec x \rightarrow f'(x) = \sec x \tan x$
- $f(x) = \cot x \rightarrow f'(x) = -\cosec^2 x$
- $f(x) = \cosec x \rightarrow f'(x) = -\cosec x \cot x$
- If $y = \dfrac{f(x)}{g(x)}$ then $\dfrac{dy}{dx} = \dfrac{f'(x)g(x) - f(x)g'(x)}{\{g(x)\}^2}$
- $f(x) = \sec^2 kx \rightarrow \int f(x)\, dx = \dfrac{1}{k} \tan kx + c$
- $f(x) = \tan x \rightarrow \int f(x)\, dx = \ln|\sec x| + c$
- $f(x) = \cot x \rightarrow \int f(x)\, dx = \ln|\sin x| + c$
- $f(x) = \cosec x \rightarrow \int f(x)\, dx = -\ln|\cosec x + \cot x| + c$
- $f(x) = \sec x \rightarrow \int f(x)\, dx = \ln|\sec x + \tan x| + c$
- $\int u \dfrac{dv}{dx}\, dx = uv - \int v \dfrac{du}{dx}\, dx$

Not all the formulae and results that you might need in C4 are included in the *List of Formulae* booklet. You must know these extra formulae and results. The following exercise invites you to see if you can remember them.

Exercise 30

(1) Given $y = \cos kx$, find $\dfrac{dy}{dx}$.

(2) Express $\sin^2 \theta$ in terms of $\cos 2\theta$.

(3) Write down the formula for the volume of the solid produced when a curve is rotated completely about the x-axis.

(4) Complete the logarithm law $\ln p - \ln q = \ldots$.

(5) Write down the equation of the circle with radius r and centre (a, b).

(6) Find $\int e^{kx}\, dx$.

(7) Write down an identity linking $\tan^2 \theta$ and $\sec^2 \theta$.

(8) Write down the formula giving the magnitude of the vector $x\mathbf{i} + y\mathbf{j} + z\mathbf{k}$.

(9) Find $\int \sin kx\, dx$.

(10) Express $\cos^2 \theta$ in terms of $\cos 2\theta$.

(11) Write down an identity linking $\cot^2 \theta$ and $\cosec^2 \theta$.

(12) Write down the condition for the equation $ax^2 + bx + c = 0$ to have two different real roots.

(13) Find $\int \dfrac{1}{ax + b}\, dx$.

(14) Find $\int \dfrac{1}{(ax + b)^2}\, dx$.

(15) Complete the identity $\sin 2\theta = \ldots$.

(16) Given $y = \sin kx$, find $\dfrac{dy}{dx}$.

(17) Convert 135° to radians.

(18) Find $\int \cos kx\, dx$.

(19) State the name given to the result $y = f(x)g(x) \rightarrow \dfrac{dy}{dx} = f'(x)g(x) + f(x)g'(x)$.

(20) For the quadratic equation $ax^2 + bx + c = 0$, write down the formula that gives the two roots of the equation.

Synoptic assessment

In mathematics, the requirement for synoptic assessment is met automatically because of the nature of the subject. In answering C4 questions, you are often using results and techniques that you met first in C1, C2 or C3.

The following exercise consists of C4 questions that depend in part on ideas from your study of earlier modules.

Exercise 31

(1) The curve $y = \sin 2x$, for $0 \leqslant x \leqslant \tfrac{1}{4}\pi$, is rotated completely about the x-axis. Find the exact volume of the solid produced.

(2) A curve passes through the point $(5, -2)$ and its equation satisfies the differential equation $\dfrac{dy}{dx} = \dfrac{1 - x}{4 + y}$.

  (i) Solve the differential equation to find the particular solution.

  (ii) Show that the curve is a circle and find its centre and radius.

(3) A curve is defined by the parametric equations $x = 4e^{2t} - 3t$, $y = 5e^{-t} + t$. Find the gradient of the normal to the curve at the point for which $t = \ln 2$.

(4) A curve is defined by the parametric equations $x = 3 \operatorname{cosec} \theta + 4$, $y = 2 \cot \theta - 5$. Find a cartesian equation of the curve.

(5) Find the binomial series for $\sqrt{\dfrac{1 - 12x}{1 + 12x}}$ up to and including the term in x^2. By substituting $x = 0.002$, find an approximation, correct to 4 decimal places, for $\sqrt{61}$.

(6) The equation of a curve is $y = e^{\frac{1}{3}x} \tan 2x$. Show that the curve has no stationary points.

A2 Mathematics

This section consists of three practice C4 examination papers. Each paper is designed to be of the same style and standard as an actual C4 examination paper. Each paper is worth a total of 72 marks and the time allowed is 1 hour 30 minutes. The best approach is to set aside this time and work steadily through a paper. Practising in this way should help significantly when you come to take the examination.

Remember the following points:
- The early questions in the paper should be straightforward, whereas later questions may contain some challenging aspects.
- The best strategy is to work through the questions in order.
- Although these are practice papers, answer them with care — you do not want to develop bad habits.
- Have the *List of Formulae* booklet to hand as you attempt these papers. It is an important aid as you answer the questions.
- In the examination, you must provide clear and full solutions, otherwise you risk losing marks. Do the same with these practice papers.
- Make sure that your calculator is in good working order. Also make sure that you have it in the correct mode; in this module, you will usually need it in RADIAN mode.

For each of the first two practice papers, there is a set of hints and suggestions, question by question. Only consult this if you are really stuck; it is much more beneficial if you can sort out problems without looking at the hints. There will not be any hints available to you in the examination and dealing with problems is an important examination technique that needs practice.

Examiner's comments

Solutions to the questions are given. These are followed by examiner's comments, indicated by the icon *e*. These point out various issues such as common errors, tricky points, alternative methods etc. Once you have completed a paper, check your solutions and read carefully through the comments.

OCR Unit 4724

Specimen paper 1

Question 1
Find the quotient and remainder when $2x^3 - 7x - 8$ is divided by $x^2 + 3x - 2$. (4 marks)

Question 2
Find:
(i) $\int 4\sec^2 2x \, dx$ (2 marks)
(ii) $\int \dfrac{\cos 2x}{1 - \sin 2x} \, dx$ (3 marks)

Question 3
A curve has parametric equations $x = 2\sin\theta + 1$, $y = 4\cos 2\theta - 1$. Find an expression, in terms of θ, for $\dfrac{dy}{dx}$ in its simplest form. (5 marks)

Question 4
Find the exact value of $\int_1^e 4x^3 \ln x \, dx$. (5 marks)

Question 5
Resolve $\dfrac{4x^2 - 14x + 20}{x^3 - 4x^2 + 4x}$ into partial fractions. (6 marks)

Question 6
Use the substitution $u = 2x + 1$ to find the exact value of $\int_0^3 \dfrac{2x}{2x + 1} \, dx$. (6 marks)

Question 7
Two straight lines have vector equations $r = \begin{pmatrix} 12 \\ -7 \\ 2 \end{pmatrix} + \lambda \begin{pmatrix} 2 \\ -1 \\ -2 \end{pmatrix}$ and $r = \begin{pmatrix} -16 \\ 2 \\ 15 \end{pmatrix} + \mu \begin{pmatrix} -6 \\ 2 \\ 3 \end{pmatrix}$.

(i) Show that the lines intersect and find the coordinates of the point of intersection. (5 marks)

(ii) Find the acute angle between the lines. (4 marks)

Question 8
An economist proposes that a model showing the value, P, of a cost of living index at a time t years from now is the differential equation $\sqrt{t+1}\,\dfrac{dP}{dt} = \dfrac{1}{10}P$. It is given that $P = 100$ when $t = 0$.

(i) Solve the differential equation to find P in terms of t. (6 marks)

(ii) Determine the predicted value of the index 12 years from now. (1 mark)

(iii) Calculate the value of t for which the predicted value of the index is 225. (2 marks)

Question 9

(i) Find the series expansion of $(1 + 4x)^{\frac{3}{2}}$ up to and including the term in x^3. (4 marks)

(ii) Hence, or otherwise, find the series expansion of $(1 - 4x)^{\frac{3}{2}}$ up to and including the term in x^3, and show that, for small values of x,

$(1 + 4x)^{\frac{3}{2}} - (1 - 4x)^{\frac{3}{2}} \approx 12x - 8x^3.$ (3 marks)

(iii) By using $x = 0.01$, find an approximate value for $26^{\frac{3}{2}} - 24^{\frac{3}{2}}$. (4 marks)

Question 10

The diagram shows the curve with equation $x^2 - 6xy + 4y^2 + 20 = 0$. At each of the points A and B on the curve, there is a stationary point. At each of the points C and D, the tangent to the curve is parallel to the y-axis.

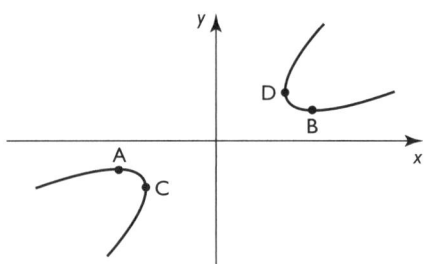

(i) Show that $\dfrac{dy}{dx} = \dfrac{x - 3y}{3x - 4y}$. (4 marks)

(ii) Find the coordinates of the points A and B. (4 marks)

(iii) Find the coordinates of the points C and D. (4 marks)

Hints and suggestions

Question 1

This is a straightforward opening question. If using a division approach, be careful to note the missing x^2 term in the expression to be divided — this means some precision is needed in setting up the division. If using an approach involving an identity, remember that the remainder should be of the form $Dx + E$.

Question 2

In an integration question, it is always worth checking a rational expression, such as the one in part (ii), to see whether it is of the form $\dfrac{kf'(x)}{f(x)}$.

Question 3

This involves a routine application of the differentiation of parametric equations but a trigonometric identity is needed in order to simplify the answer.

Question 4

Integration by parts is needed here but be sure to choose correctly for u and $\frac{dv}{dx}$. The formula is in the *List of Formulae* booklet if you are not sure.

Question 5

The first step is to factorise the denominator. This leads to a repeated factor, and this has implications for the form of the partial fractions to be set up.

Question 6

This is a standard question involving integration by substitution — don't forget to deal with the 'dx', the '$2x$' in the numerator and the limits. Note too that an exact answer is required.

Question 7

Be sure to adopt a clear process for part (i) — solving two of the equations and then checking in the third. The scalar product is needed in part (ii) — be sure to choose the correct two vectors.

Question 8

Separate the variables with care. The move from $\ln P = \ldots$ to $P = \ldots$ also needs to be handled carefully.

Question 9

The given answer in part (ii) gives you the opportunity to go back and check for errors if you do not genuinely reach $12x - 8x^3$. Part (iii) is challenging because it may not be immediately obvious how 26 and 24 are related to what has already been found. The advice in such cases is to proceed as indicated in the question, substituting $x = 0.01$ in this case, and see where that takes you.

Question 10

Part (i) is routine work with an implicitly defined equation, but do not forget to use the product rule when appropriate. For part (ii), $\frac{dy}{dx} = 0$ is clearly needed but this must be used in conjunction with the original equation. A tangent parallel to the y-axis has 'infinite' gradient and so you will need to make use of the denominator of the expression for $\frac{dy}{dx}$.

A2 Mathematics

Solutions

Question 1

$$\begin{array}{r}2x-6\\x^2+3x-2\overline{\smash{\big)}\,2x^3-7x-8}\\\underline{2x^3+6x^2-4x}\\-6x^2-3x-8\\\underline{-6x^2-18x+12}\\15x-20\end{array}$$

Quotient = $2x - 6$, remainder = $15x - 20$.

💡 Most candidates have no difficulty in making a successful start to the paper. A common mistake is to think that the remainder is a constant.

Question 2

(i) $\int 4\sec^2 2x\, dx = 2\tan 2x + c$

(ii) $\int \dfrac{\cos 2x}{1 - \sin 2x}\, dx = -\dfrac{1}{2}\int \dfrac{-2\cos 2x}{1 - \sin 2x}\, dx = -\dfrac{1}{2}\ln|1 - \sin 2x| + c$

💡 The integral in part (i) should be recognised as one in which the answer can be written down immediately. Some candidates, though, try the use of various trigonometric identities and make no progress. The integral in part (ii) is usually recognised as one involving a natural logarithm but there are errors with the constant factor and $-2\ln(1 - \sin 2x)$ is a common, incorrect, response.

Question 3

$$\dfrac{dy}{dx} = \dfrac{-8\sin 2\theta}{2\cos\theta} = \dfrac{-16\sin\theta\cos\theta}{2\cos\theta} = -8\sin\theta$$

💡 Candidates are expected to realise that the expression for $\dfrac{dy}{dx}$ can be simplified beyond $\dfrac{-4\sin 2\theta}{\cos\theta}$. This requires a familiarity with the identity for $\sin 2\theta$ and this is not shown by all candidates.

Question 4

$$\int_1^e 4x^3 \ln x\, dx = \ln x \cdot x^4 - \int_1^e \dfrac{1}{x} \cdot x^4\, dx = x^4 \ln x - \int_1^e x^3\, dx$$

$$= \left[x^4 \ln x - \dfrac{1}{4}x^4\right]_1^e = \left(e^4 \ln e - \dfrac{1}{4}e^4\right) - \left(0 - \dfrac{1}{4}\right) = \dfrac{3}{4}e^4 + \dfrac{1}{4}$$

💡 Candidates recognise this as an example requiring integration by parts. Not all make the appropriate choice for u and $\dfrac{dv}{dx}$ and some proceed with the mistaken belief that the integral of ln x is $\frac{1}{x}$. A crucial step involves $\int \frac{1}{x} \cdot x^4 \, dx$ and some candidates do not appreciate the need to simplify this to $\int x^3 \, dx$ before integrating. Many candidates, however, are able to answer this question correctly.

Question 5

Let $\dfrac{4x^2 - 14x + 20}{x(x-2)^2} \equiv \dfrac{A}{x} + \dfrac{B}{x-2} + \dfrac{C}{(x-2)^2}$

Hence $4x^2 - 14x + 20 \equiv A(x-2)^2 + Bx(x-2) + Cx$

Putting $x = 0$ gives $20 = A \times 4$, giving $A = 5$

Putting $x = 2$ gives $8 = C \times 2$, giving $C = 4$

Comparing coefficients of x^2 gives $4 = A + B$, giving $B = -1$

Hence $\dfrac{4x^2 - 14x + 20}{x^3 - 4x^2 + 4x} \equiv \dfrac{5}{x} - \dfrac{1}{x-2} + \dfrac{4}{(x-2)^2}$

Questions on partial fractions are usually answered well but this one is slightly more awkward. A minority of candidates try to proceed without a factorised denominator and a few factorise the denominator incorrectly. The form for the partial fractions in a case when a repeated factor is involved is not known by all.

Question 6

Differentiating $u = 2x + 1$ gives $\dfrac{du}{dx} = 2$ and therefore $dx = \frac{1}{2} du$

Using $u = 2x + 1$, the x limits of 0 and 3 become u limits of 1 and 7

Substituting: $\displaystyle\int_0^3 \dfrac{2x}{2x+1} \, dx = \int_1^7 \dfrac{u-1}{u} \cdot \dfrac{1}{2} \, du = \int_1^7 \left(\dfrac{1}{2} - \dfrac{1}{2u}\right) du$

$= \left[\dfrac{1}{2}u - \dfrac{1}{2}\ln u\right]_1^7 = \left(\dfrac{7}{2} - \dfrac{1}{2}\ln 7\right) - \left(\dfrac{1}{2} - 0\right) = 3 - \dfrac{1}{2}\ln 7$

Most candidates answer this question well. There are various steps, each of which some do get wrong — not dealing with the $2x$ in the numerator, forgetting to change the limits, not realising that $\dfrac{u-1}{u}$ has to be rewritten as $1 - \dfrac{1}{u}$ and, commonly, going wrong with the integral of $\dfrac{1}{2u}$.

Question 7

(i) Forming equations from the x, y and z components: $12 + 2\lambda = -16 - 6\mu$
$-7 - \lambda = 2 + 2\mu$
$2 - 2\lambda = 15 + 3\mu$

Solving the first and third equations gives $\lambda = 1$, $\mu = -5$

Checking in the second equation: left-hand side $= -7 - 1 = -8$
right-hand side $= 2 - 10 = -8$

The second equation is satisfied so the lines do meet; the point of intersection is $(14, -8, 0)$

(ii) $\begin{pmatrix} 2 \\ -1 \\ -2 \end{pmatrix} \cdot \begin{pmatrix} -6 \\ 2 \\ 3 \end{pmatrix} = -12 - 2 - 6 = -20 = \sqrt{9}\sqrt{49}\cos\theta$

Acute angle is given by $\cos\theta = \frac{20}{21}$ and hence the angle is 17.8°.

e This is a routine question on the principal aspects of vectors in this module and candidates with a sound grasp of the procedures have no difficulty in achieving high marks. It is important to have a set process for confirming that the lines meet, i.e. choosing two of the equations for λ and μ, solving them and, only then, checking that the other equation is satisfied by the values found. Some candidates adopt muddled approaches which use all three equations at different stages to find λ and μ; there is then nowhere to carry out a check.

Question 8

(i) Separating the variables: $\int \frac{1}{P}dP = \int \frac{1}{10}(t+1)^{-\frac{1}{2}}dt$

Integrating: $\ln P = \frac{1}{5}(t+1)^{\frac{1}{2}} + c$

When $t = 0$, $P = 100$, giving $\ln 100 = \frac{1}{5} + c$ and so $c = \ln 100 - \frac{1}{5}$

Particular solution is: $\ln P = \frac{1}{5}(t+1)^{\frac{1}{2}} + \ln 100 - \frac{1}{5}$, leading to $P = e^{\frac{1}{5}(t+1)^{\frac{1}{2}}} \cdot 100 \cdot e^{-\frac{1}{5}}$

Hence $P = 100 e^{\frac{1}{5}\sqrt{t+1} - \frac{1}{5}}$ or $P = 81.9 e^{\frac{1}{5}\sqrt{t+1}}$

(ii) Substituting $t = 12$: $P = 100 e^{\frac{1}{5}\sqrt{13} - \frac{1}{5}} = 168$

So predicted value of the index is 168

(iii) Substituting $P = 225$: $225 = 100 e^{\frac{1}{5}\sqrt{t+1} - \frac{1}{5}}$ giving $\frac{1}{5}\sqrt{t+1} - \frac{1}{5} = \ln 2.25$ and therefore $t = 24.5$

e This question is set in a context and some candidates find that the context obscures the mathematical steps to be taken. In fact, the solution of the differential equation is straightforward, although finding an explicit expression for P needs some careful manipulation. Parts (ii) and (iii) assess whether candidates can relate the mathematics to the context but neither request is difficult. Candidates earning full marks on this question are showing work of grade-B standard or better.

Question 9

(i) $(1 + 4x)^{\frac{3}{2}} = 1 + \frac{3}{2} \cdot 4x + \frac{\frac{3}{2} \cdot \frac{1}{2}}{2} \cdot 16x^2 + \frac{\frac{3}{2} \cdot \frac{1}{2} \cdot -\frac{1}{2}}{6} \cdot 64x^3 = 1 + 6x + 6x^2 - 4x^3$

(ii) $(1 - 4x)^{\frac{3}{2}} = 1 - 6x + 6x^2 + 4x^3$

Hence $(1 + 4x)^{\frac{3}{2}} - (1 - 4x)^{\frac{3}{2}} \approx (1 + 6x + 6x^2 - 4x^3) - (1 - 6x + 6x^2 + 4x^3)$

giving $(1 + 4x)^{\frac{3}{2}} - (1 - 4x)^{\frac{3}{2}} \approx 12x - 8x^3$

(iii) Substituting $x = 0.01$ gives $1.04^{\frac{3}{2}} - 0.96^{\frac{3}{2}} \approx 0.119\,992$

$\left(\dfrac{26}{25}\right)^{\frac{3}{2}} - \left(\dfrac{24}{25}\right)^{\frac{3}{2}} \approx 0.119\,992$

$\dfrac{26^{\frac{3}{2}}}{125} - \dfrac{24^{\frac{3}{2}}}{125} \approx 0.119\,992$

and hence $26^{\frac{3}{2}} - 24^{\frac{3}{2}} \approx 14.999$

e The expansion in part (i) is easily produced although, for some candidates, lack of attention to detail leads to errors. The expansion in part (ii) can then be written down by changing the sign of each term involving an odd power of x; however, many candidates prefer to start afresh in finding the second expansion. Examiners expect to see plenty of detail as candidates confirm the given result. Part (iii) is challenging because of the need to see how 26 and 24 can be obtained; candidates succeeding here are exhibiting work of grade-A standard.

Question 10

(i) Differentiating, $2x - 6y - 6x\dfrac{dy}{dx} + 8y\dfrac{dy}{dx} + 0 = 0$, giving $2x - 6y = \dfrac{dy}{dx}(6x - 8y)$

and hence $\dfrac{dy}{dx} = \dfrac{2x - 6y}{6x - 8y} = \dfrac{x - 3y}{3x - 4y}$

(ii) Stationary points are given by $\dfrac{dy}{dx} = 0$, i.e. $x - 3y = 0$

Substituting $x = 3y$ into the original equation: $9y^2 - 6 \cdot 3y \cdot y + 4y^2 + 20 = 0$

giving $5y^2 = 20$ and therefore $y = \pm 2$

Using $x = 3y$, $x = \pm 6$ and so A is $(-6, -2)$ and B is $(6, 2)$

(iii) At C and D, $\dfrac{dy}{dx}$ is infinite, i.e. $3x - 4y = 0$

Substituting $x = \dfrac{4}{3}y$ into the original equation: $\dfrac{16}{9}y^2 - 6 \cdot \dfrac{4}{3}y \cdot y + 4y^2 + 20 = 0$

giving $\dfrac{20}{9}y^2 = 20$ and therefore $y = \pm 3$

Using $x = \dfrac{4}{3}y$, $x = \pm 4$ and so C is $(-4, -3)$ and D is $(4, 3)$

e The curve is an unfamiliar one but differentiating the implicitly defined equation is a familiar process to most candidates. In part (ii), many candidates realise that $x - 3y = 0$ but are not sure what to do thereafter. Part (iii) is more challenging and many do not appreciate the need for $3x - 4y$ to be zero.

Specimen paper 2

Question 1
The expressions $p(x)$ and $q(x)$ are defined by $p(x) = \dfrac{x+3}{2x-2}$ and $q(x) = \dfrac{x^2+6x+9}{x^2-2x+1}$.

Express $\dfrac{q(x)}{p(x)}$ in simplified form. (4 marks)

Question 2
A curve has equation $x^3 + y \ln y = 8$. Find the gradient of the curve at the point $(2, 1)$.

(5 marks)

Question 3
Differentiate with respect to x:
(i) $x^2 \sin 4x$ (3 marks)
(ii) $\dfrac{\tan 3x}{x+1}$ (3 marks)

Question 4
Find the exact value of $\displaystyle\int_{\frac{1}{8}\pi}^{\frac{1}{4}\pi} 6 \cos^2 2x \, dx$. (6 marks)

Question 5
(i) Expand $\dfrac{1}{(2+3x)^2}$ in ascending powers of x up to and including the term in x^2. (4 marks)

(ii) Hence determine the coefficient of x^2 in the expansion of $\dfrac{8-3x}{(2+3x)^2}$. (3 marks)

Question 6
It is given that $f(x) = \dfrac{3-x}{(1+3x)(2+x)}$. Resolve $f(x)$ into partial fractions and hence show that $\displaystyle\int_0^{21} f(x)\, dx = \ln\dfrac{32}{23}$. (7 marks)

Question 7
Find $\displaystyle\int 54x^2 e^{3x}\, dx$. (7 marks)

Question 8
(i) Find the general solution of the differential equation $\dfrac{dy}{dx} = 4x \cos^2 2y$. (4 marks)

(ii) Find the particular solution of the differential equation for which $y = \dfrac{1}{8}\pi$ when $x = 2$, giving your answer in the form $y = f(x)$. (4 marks)

Question 9
Points A and B have coordinates $(3, 11, 2)$ and $(7, 1, -2)$ respectively. The line l passes through B and is parallel to the vector $\mathbf{i} + 2\mathbf{j} - 4\mathbf{k}$. The line l meets the x-axis at the point C.

(i) Write down a vector equation for the line *l*. (2 marks)
(ii) Find the coordinates of C. (3 marks)
(iii) Show that AB is perpendicular to the line *l* and hence find the area of the triangle ABC, giving your answer correct to 3 significant figures. (5 marks)

Question 10

A curve has parametric equations $x = t^3 + 3t$, $y = t^4 - 4t^2$.
The diagram shows the curve which meets the x-axis at the origin and at the points P and Q.

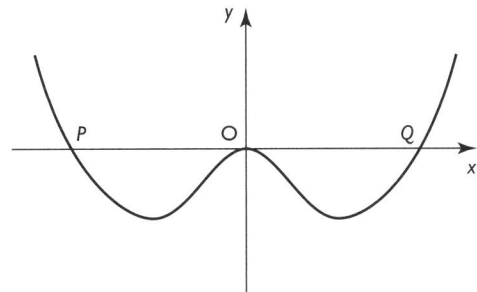

(i) Find the exact coordinates of the stationary points of the curve. (7 marks)
(ii) The tangents to the curve at the points P and Q meet each other at the point R. Find the exact coordinates of R. (5 marks)

Hints and suggestions

Question 1

The first step is to factorise the various expressions including $2x - 2$. Expect that some cancellation will be possible when you have assembled the complete expression.

Question 2

This requires the differentiation (with respect to x) of an implicitly defined equation. Note that the product rule is needed and remember that the right-hand side must be differentiated too.

Question 3

This is a straightforward application of the two differentiation rules. The quotient rule is in the *List of Formulae* booklet.

Question 4

This is a typical integration requiring the use of an identity to convert $\cos^2 2x$ into an expression involving $\cos 4x$. Note that an exact value is required.

A2 Mathematics

Question 5
Care is needed to convert $\dfrac{1}{(2+3x)^2}$ to the form $k\left(1+\dfrac{3}{2}x\right)^{-2}$ before attempting the expansion. Further precise work is needed with the expansion itself.

Question 6
With the answer given, you will need to provide accurate, detailed work to present a solution that is convincing enough to earn all the marks. Don't make the mistake of thinking that the integral of $\dfrac{1}{1+3x}$ is $\ln|1+3x|$.

Question 7
This requires integration by parts — twice.

Question 8
Separating the variables will lead to $\int \dfrac{1}{\cos^2 2y}\,dy$; note that this is $\int \sec^2 2y\,dy$ and integration is now possible. Don't forget the '... + c' in the general solution. Without this constant in the general solution there is nothing to do in part (ii).

Question 9
Note that, in part (ii), for any point on the x-axis, $y = z = 0$. Remember that, in part (iii), perpendicularity is demonstrated by showing that the scalar product is zero.

Question 10
You will need to find the values of the parameter t at the points P and Q and these correspond to $y = 0$.

Solutions

Question 1
Factorising: $p(x) = \dfrac{x+3}{2(x-1)}$, $q(x) = \dfrac{(x+3)^2}{(x-1)^2}$. Hence $\dfrac{q(x)}{p(x)} = \dfrac{(x+3)^2}{(x-1)^2} \cdot \dfrac{2(x-1)}{x+3} = \dfrac{2(x+3)}{x-1}$

✎ Most candidates find this a straightforward first question. Some show confusion when substituting for $p(x)$ and put the factors in incorrect places.

Question 2
Differentiating with respect to x: $3x^2 + \dfrac{dy}{dx} \cdot \ln y + y \cdot \dfrac{1}{y} \cdot \dfrac{dy}{dx} = 0$

$$3x^2 + \dfrac{dy}{dx}(\ln y + 1) = 0$$

OCR Unit 4724

Substituting $x = 2, y = 1$: $12 + \frac{dy}{dx} \cdot 1 = 0$ giving $\frac{dy}{dx} = -12$, i.e. the gradient of the curve is -12.

e Not all candidates recognise the need for the product rule and, after differentiation, 8 is sometimes retained on the right-hand side. For candidates with a sure grasp of dealing with this topic, this question presents no problems.

Question 3

(i) $2x \sin 4x + 4x^2 \cos 4x$

(ii) $\dfrac{(x+1) \cdot 3 \sec^2 3x - \tan 3x \cdot 1}{(x+1)^2} = \dfrac{3(x+1) \sec^2 3x - \tan 3x}{(x+1)^2}$

e Almost all candidates are able to answer both parts of this question with no difficulty, showing a sound grasp of the differentiation techniques involved.

Question 4

$$\text{Integral} = \int_{\frac{1}{8}\pi}^{\frac{1}{4}\pi} 6\left(\frac{1}{2} + \frac{1}{2}\cos 4x\right) dx = \int_{\frac{1}{8}\pi}^{\frac{1}{4}\pi} (3 + 3\cos 4x)\, dx$$

$$= \left[3x + \frac{3}{4}\sin 4x\right]_{\frac{1}{8}\pi}^{\frac{1}{4}\pi} = \frac{3}{8}\pi - \frac{3}{4}$$

e This question provides a good test of candidates' ability with integration and trigonometry. Not all candidates appreciate the need to convert $6\cos^2 2x$ to a form that can be integrated and incorrect integrals such as $2\cos^3 2x$ are not uncommon. An accurate solution producing the required exact answer indicates work of grade-C standard or better.

Question 5

(i) $\dfrac{1}{(2+3x)^2} = \dfrac{1}{4}\left(1 + \dfrac{3}{2}x\right)^{-2} = \dfrac{1}{4}\left(1 + (-2) \cdot \dfrac{3}{2}x + \dfrac{-2 \cdot -3}{2}\left(\dfrac{3}{2}x\right)^2\right)$

$$= \frac{1}{4} - \frac{3}{4}x + \frac{27}{16}x^2$$

(ii) $\dfrac{8-3x}{(2+3x)^2} = (8-3x)\left(\dfrac{1}{4} - \dfrac{3}{4}x + \dfrac{27}{16}x^2 + \ldots\right)$

Coefficient of $x^2 = 8 \times \dfrac{27}{16} + 3 \times \dfrac{3}{4} = \dfrac{63}{4}$

e The first step is the most challenging part of this question. Most candidates know that the form $(1 + \ldots)^{-2}$ is involved but many are unsure with the necessary algebraic manipulation, and errors such as $2\left(1 + \dfrac{3}{2}x\right)^{-2}$ are common. The procedure for finding the binomial series is usually carried out well and, even if candidates have gone wrong with the first step, several marks can still be earned. Part (ii) is usually handled well although some candidates do more than is necessary by carrying out a complete expansion up to the x^2 term.

A2 Mathematics

Question 6

Let $\dfrac{3-x}{(1+3x)(2+x)} \equiv \dfrac{A}{1+3x} + \dfrac{B}{2+x}$, giving $3 - x \equiv A(2+x) + B(1+3x)$

Substituting $x = -2$ gives $5 = B \times (-5)$, giving $B = -1$

Substituting $x = -\dfrac{1}{3}$ gives $\dfrac{10}{3} = A \times \dfrac{5}{3}$, giving $A = 2$

Hence $f(x) \equiv \dfrac{2}{1+3x} - \dfrac{1}{2+x}$

$$\int_0^{21} f(x)\,dx = \int_0^{21}\left(\dfrac{2}{1+3x} - \dfrac{1}{2+x}\right)dx = \left[\dfrac{2}{3}\ln|1+3x| - \ln|2+x|\right]_0^{21}$$

$$= \left(\dfrac{2}{3}\ln 64 - \ln 23\right) - \left(\dfrac{2}{3}\ln 1 - \ln 2\right)$$

$$= \ln 64^{\frac{2}{3}} - \ln 23 + \ln 2$$

$$= \ln 16 - \ln 23 + \ln 2 = \ln \dfrac{16 \times 2}{23} = \ln \dfrac{32}{23}$$

e The resolution into partial fractions is generally handled well and only careless slips prevent a correct answer being obtained. Work with the integration is not always so assured. A common error is to integrate $\dfrac{2}{1+3x}$ as $2\ln(1+3x)$. With the answer given, a thorough solution is required and many candidates do not provide sufficient detail to be fully convincing. In cases where the answer $\ln \dfrac{32}{23}$ is not genuinely obtained, there is little evidence of trying to find the earlier error or errors.

Question 7

$$\int 54x^2 e^{3x}\,dx = 54x^2 \cdot \dfrac{1}{3}e^{3x} - \int 108x \cdot \dfrac{1}{3}e^{3x}\,dx = 18x^2 e^{3x} - \int 36xe^{3x}\,dx$$

$$= 18x^2 e^{3x} - \left(36x \cdot \dfrac{1}{3}e^{3x} - \int 36 \cdot \dfrac{1}{3}e^{3x}\,dx\right)$$

$$= 18x^2 e^{3x} - 12xe^{3x} + \int 12e^{3x}\,dx$$

$$= 18x^2 e^{3x} - 12xe^{3x} + 4e^{3x} + c$$

e Most candidates recognise this as a question requiring integration by parts, but many do not also attempt integration by parts at the second stage when faced with $\int 36xe^{3x}\,dx$. Candidates who set out their solutions meticulously and with due attention to detail benefit here; they do not make sign slips and other careless errors. A successful outcome suggests work of at least grade-B standard.

Question 8

(i) Separating the variables: $\int \dfrac{1}{\cos^2 2y}\,dy = \int 4x\,dx$, leading to $\int \sec^2 2y\,dy = \int 4x\,dx$

Integrating: $\dfrac{1}{2}\tan 2y = 2x^2 + c$

General solution is $\tan 2y = 4x^2 + C$ where $C = 2c$

(ii) Substituting $x = 2, y = \frac{1}{8}\pi$: $\tan \frac{1}{4}\pi = 16 + C$, giving $C = -15$

Particular solution is $\tan 2y = 4x^2 - 15$

i.e. $2y = \tan^{-1}(4x^2 - 15)$ and therefore $y = \frac{1}{2}\tan^{-1}(4x^2 - 15)$

🖉 This is not a difficult differential equation but a crucial step is recognising that $\frac{1}{\cos^2 2y}$ is the same as $\sec^2 2y$. Only candidates with a sure grasp of the integration of trigonometric functions realise that this is a significant step. Not all candidates are confident with inverse trigonometric functions and the step from $\tan 2y = \ldots$ to $y = \ldots$ is awkward for some.

Question 9

(i) $\mathbf{r} = \begin{pmatrix} 7 \\ 1 \\ -2 \end{pmatrix} + \lambda \begin{pmatrix} 1 \\ 2 \\ -4 \end{pmatrix}$

(ii) l meets the x-axis where $y = z = 0$, i.e. where $\lambda = -\frac{1}{2}$ giving $(6\frac{1}{2}, 0, 0)$

(iii) $\overrightarrow{AB} = \begin{pmatrix} 7 \\ 1 \\ -2 \end{pmatrix} - \begin{pmatrix} 3 \\ 11 \\ 2 \end{pmatrix} = \begin{pmatrix} 4 \\ -10 \\ -4 \end{pmatrix}$ and the direction of $l = \begin{pmatrix} 1 \\ 2 \\ -4 \end{pmatrix}$

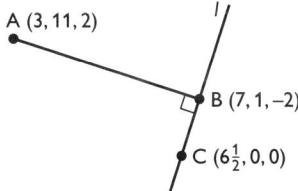

A (3, 11, 2)

B (7, 1, −2)

C $(6\frac{1}{2}, 0, 0)$

Forming the scalar product: $\begin{pmatrix} 4 \\ -10 \\ -4 \end{pmatrix} \cdot \begin{pmatrix} 1 \\ 2 \\ -4 \end{pmatrix} = 4 - 20 + 16 = 0$

The scalar product is zero, so AB is perpendicular to l.

$|\overrightarrow{AB}| = \sqrt{16 + 100 + 16} = \sqrt{132}$ and $|\overrightarrow{BC}| = \sqrt{\frac{1}{4} + 1 + 4} = \sqrt{5\frac{1}{4}}$

Area of triangle ABC $= \frac{1}{2} \cdot \sqrt{132} \cdot \sqrt{5\frac{1}{4}} = 13.2$ square units

🖉 This question provides a test of various vector techniques. The only tricky part is finding the coordinates of C; not all candidates are confident with geometrical ideas in three dimensions and so do not readily appreciate that a point on the x-axis must have zero y- and z-coordinates. A simple diagram is a help with part (iii) but most candidates proceed without and some confusion is apparent.

Question 10

(i) $\frac{dy}{dx} = \frac{4t^3 - 8t}{3t^2 + 3}$

For stationary points, $\dfrac{dy}{dx} = 0$ giving $4t(t^2 - 2) = 0$ and hence $t = 0, \pm\sqrt{2}$

$t = 0$ gives $(0, 0)$

$t = -\sqrt{2}$ gives $x = -2\sqrt{2} - 3\sqrt{2} = -5\sqrt{2}$ and $y = 4 - 8 = -4$

$t = \sqrt{2}$ gives $x = 2\sqrt{2} + 3\sqrt{2} = 5\sqrt{2}$ and $y = 4 - 8 = -4$

The stationary points are $(0, 0)$, $(-5\sqrt{2}, -4)$ and $(5\sqrt{2}, -4)$

(ii) P and Q are given by $y = 0$, i.e. $t^4 - 4t^2 = 0$, giving $t = 0, -2, 2$

Hence $t = -2$ at P and P is $(-14, 0)$ and $t = 2$ at Q and Q is $(14, 0)$

At P, $\dfrac{dy}{dx} = \dfrac{-32 + 16}{15} = -\dfrac{16}{15}$ and the tangent is $y = -\dfrac{16}{15}(x + 14)$

At Q, $\dfrac{dy}{dx} = \dfrac{32 - 16}{15} = \dfrac{16}{15}$ and the tangent is $y = \dfrac{16}{15}(x - 14)$

The tangents meet where $-\dfrac{16}{15}(x + 14) = \dfrac{16}{15}(x - 14)$, giving $x = 0$ and $y = -\dfrac{224}{15}$

So the coordinates of R are $\left(0, -\dfrac{224}{15}\right)$

The concept of parametric equations — understanding how parameter values, coordinates and the sketch are related — eludes some candidates. The third variable seems to confuse them and working with a cartesian equation is preferred. That is not an option here and it must be values of the parameter t that are established for P, Q and the stationary points. Candidates showing assurance and accuracy in answering both parts of this question are demonstrating work at grade-A standard.

Specimen paper 3

Question 1
Three points P, Q and R have coordinates $P(4,-2,-7)$, $Q(1,0,5)$ and $R(-2,2,3)$. Find the angle QPR. (5 marks)

Question 2
(i) Find the quotient and remainder when $x^4 - 3x^3 + 5x^2 - x + 6$ is divided by $x^2 + x - 4$. (4 marks)

(ii) Deduce, or find otherwise, the remainder when $x^4 - 3x^3 + 5x^2 + 29x + 6$ is divided by $x^2 + x - 4$. (2 marks)

Question 3
Use integration by parts to find the exact value of $\int_1^2 \ln(3x)\,dx$. (6 marks)

Question 4
A curve has equation $4x^2 + y^2 + 8\sin\tfrac{1}{2}y = 16$. Find the equation of the normal to the curve at the point $(2, 0)$. (6 marks)

Question 5
The vector equations of two straight lines are $r = \begin{pmatrix} 2 \\ -5 \\ 3a \end{pmatrix} + \lambda \begin{pmatrix} 1 \\ 4 \\ -a \end{pmatrix}$ and $r = \begin{pmatrix} 1 \\ 35 \\ 2 \end{pmatrix} + \mu \begin{pmatrix} -2 \\ 3 \\ a \end{pmatrix}$, where a is a constant.

(i) Show that, for all values of a, the two lines are skew. (5 marks)

(ii) Given that the two lines are perpendicular, find the possible values of a. (3 marks)

Question 6
A curve has parametric equations $x = \dfrac{t^2 + 2}{t^2 - 1}$, $y = 2t^3 + 1$. Find the equation of the tangent to the curve at the point $(2, -15)$. (8 marks)

Question 7
(a) Use the substitution $u = \cos x$ to find $\int \sin^3 x \, dx$. (4 marks)

(b) Use the substitution $u = \tan 2x$ to find the exact value of $\int_0^{\frac{1}{6}\pi} \tan^3 2x \sec^4 2x \, dx$. (6 marks)

Question 8
It is given that $f(x) = \dfrac{7 - 2x - 29x^2}{(1 - x)(2 + x)(1 - 3x)}$.

(i) Express $f(x)$ in partial fractions. (5 marks)

(ii) Hence find the binomial expansion of $f(x)$ up to and including the term in x^2. (5 marks)

(iii) State the set of values of x for which the expansion in part (ii) is valid. (2 marks)

A2 Mathematics

Question 9
A lawn is infested with moss and, in an attempt to remove the moss, a course of treatment is started. At the time the treatment is started, the area of lawn infested with moss is 60 m². The area infested at a time t years after the start of the treatment is A m². A model linking the variables A and t is the differential equation $\frac{dA}{dt} = -0.2t(A-4)$.

(i) Find an expression for A in terms of t. (7 marks)
(ii) Find the value of t for which $A = 10$. (3 marks)
(iii) According to the model, what happens in the long term? (1 mark)

Solutions

Question 1

$$\vec{PQ} = \begin{pmatrix} 1 \\ 0 \\ 5 \end{pmatrix} - \begin{pmatrix} 4 \\ -2 \\ -7 \end{pmatrix} = \begin{pmatrix} -3 \\ 2 \\ 12 \end{pmatrix}; \vec{PR} = \begin{pmatrix} -2 \\ 2 \\ 3 \end{pmatrix} - \begin{pmatrix} 4 \\ -2 \\ -7 \end{pmatrix} = \begin{pmatrix} -6 \\ 4 \\ 10 \end{pmatrix}$$

Using the scalar product: $\begin{pmatrix} -3 \\ 2 \\ 12 \end{pmatrix} \cdot \begin{pmatrix} -6 \\ 4 \\ 10 \end{pmatrix} = 18 + 8 + 120 = \sqrt{157}\sqrt{152} \cos P$

Hence $\cos P = \dfrac{146}{\sqrt{157}\sqrt{152}}$, giving $QPR = 19.1°$

⚠ Care is needed to find the correct angle; this means using the scalar product $\vec{PQ} \cdot \vec{PR}$ (or $\vec{QP} \cdot \vec{RP}$). Some candidates use the scalar product with two vectors apparently chosen at random.

Question 2
(i)
```
                    x² − 4x  + 13
       _____
x² + x − 4 ) x⁴ − 3x³ + 5x² − x   + 6
             x⁴ +  x³ − 4x²
             _____
                 − 4x³ + 9x² − x
                 − 4x³ − 4x² + 16x
                 _____
                        13x² − 17x + 6
                        13x² + 13x − 52
                        _____
                              − 30x + 58
```

Hence the quotient $= x^2 - 4x + 13$ and the remainder $= -30x + 58$

(ii) The quartic expression from part (i) has increased by $30x$. So the remainder will also increase by $30x$. Hence the remainder $= 58$

⚠ Candidates adopting a long-division process generally answer this question accurately and sign slips are the only major cause of errors. Remember that when dividing by a quadratic expression, the remainder must be of the form $Dx + E$. A moment's thought enables the answer to part (ii) to be written down but many candidates prefer to carry out the whole process a second time.

Question 3

$\int \ln(3x)\,dx = \int \ln(3x) \cdot 1\,dx = \ln(3x) \cdot x - \int \frac{3}{3x} \cdot x\,dx = x\ln(3x) - \int 1\,dx = x\ln(3x) - x + c$

Hence $\int_1^2 \ln(3x)\,dx = \left[x\ln(3x) - x\right]_1^2 = (2\ln 6 - 2) - (1\ln 3 - 1)$

$= \ln 36 - 2 - \ln 3 + 1 = \ln 12 - 1$

e Writing the integral as $\int \ln(3x) \cdot 1\,dx$ is the key and, although candidates have presumably met a similar process for finding $\int \ln x\,dx$, many do not readily adopt the approach in this case. A further difficulty concerns the derivative of $\ln(3x)$, which for many candidates is $\frac{1}{3x}$. An accurate solution here indicates work of grade-C standard or better.

Question 4

Differentiating with respect to x: $8x + 2y\frac{dy}{dx} + 4\cos\frac{1}{2}y \cdot \frac{dy}{dx} = 0$

Substituting $x = 2, y = 0$: $16 + 0 + 4\frac{dy}{dx} = 0$, giving $\frac{dy}{dx} = -4$

Gradient of the normal = $\frac{1}{4}$

Equation of the normal is $y - 0 = \frac{1}{4}(x - 2)$ or $y = \frac{1}{4}x - \frac{1}{2}$

e This is a straightforward example involving an implicitly defined equation, and most candidates are able to proceed correctly. A few reveal uncertainty about the definition of the normal to a curve.

Question 5

(i) Assuming that lines meet, equating x, y and z gives: $2 + \lambda = 1 - 2\mu$

$-5 + 4\lambda = 35 + 3\mu$

$3a - a\lambda = 2 + a\mu$

Solving the first two equations gives $\lambda = 7, \mu = -4$
Checking in the third equation: left-hand side = $3a - 7a = -4a$; right-hand side = $2 - 4a$.
$-4a$ cannot equal $2 - 4a$ and so the three equations cannot be solved; there is no point of intersection and the lines are skew for all values of a.

(ii) $\begin{pmatrix} 1 \\ 4 \\ -a \end{pmatrix} \cdot \begin{pmatrix} -2 \\ 3 \\ a \end{pmatrix} = 0$, giving $-2 + 12 - a^2 = 0$ and hence $a = \pm\sqrt{10}$

e The presence of the constant a makes this a more challenging question and, although candidates generally know how to proceed with part (i), convincing solutions are not that common. Work on part (ii) is better, although some candidates use the wrong vectors. Sound work on this question indicates grade-B standard or better.

A2 Mathematics

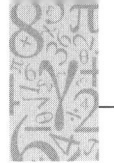

Question 6

Differentiating: $\dfrac{dx}{dt} = \dfrac{(t^2 - 1)2t - (t^2 + 2)2t}{(t^2 - 1)^2} = \dfrac{-6t}{(t^2 - 1)^2}$ and $\dfrac{dy}{dt} = 6t^2$

Hence $\dfrac{dy}{dx} = \dfrac{6t^2(t^2 - 1)^2}{-6t} = -t(t^2 - 1)^2$

The value of the parameter at $(2, -15)$ is given by $2t^3 + 1 = -15$ and hence $t = -2$
Gradient of the tangent $= -(-2) \cdot 3^2 = 18$
So the equation of the tangent is $y + 15 = 18(x - 2)$ or $y = 18x - 51$

🅔 Candidates generally understand how to produce an expression for $\dfrac{dy}{dx}$ in terms of t but accurate use of the quotient rule is sometimes lacking. The value of the parameter corresponding to $(2, -15)$ must be found. This is easily done if the y-coordinate is used but some candidates use the x-coordinate and decide incorrectly that t is 2.

Question 7

(a) Differentiating $u = \cos x$ gives $\dfrac{du}{dx} = -\sin x$ and therefore $dx = \dfrac{-du}{\sin x}$

Substituting: $\int \sin^3 x \, dx = \int \sin^3 x \cdot \dfrac{-du}{\sin x} = \int -\sin^2 x \, du = \int -(1 - u^2) \, du = \int (u^2 - 1) \, du$

$= \dfrac{1}{3}u^3 - u + c = \dfrac{1}{3}\cos^3 x - \cos x + c$

(b) Differentiating $u = \tan 2x$ gives $\dfrac{du}{dx} = 2\sec^2 2x$ and therefore $dx = \dfrac{du}{2\sec^2 2x}$

Using $u = \tan 2x$, the x limits 0 and $\dfrac{1}{6}\pi$ become u limits 0 and $\sqrt{3}$.

Substituting: $\displaystyle\int_0^{\frac{1}{6}\pi} \tan^3 2x \sec^4 2x \, dx = \int_0^{\sqrt{3}} u^3 \cdot \sec^4 2x \cdot \dfrac{du}{2\sec^2 2x}$

$= \displaystyle\int_0^{\sqrt{3}} \dfrac{1}{2} u^3 \sec^2 2x \, du = \int_0^{\sqrt{3}} \dfrac{1}{2} u^3 \cdot (1 + u^2) \, du$

$= \displaystyle\int_0^{\sqrt{3}} \left(\dfrac{1}{2}u^3 + \dfrac{1}{2}u^5\right) du = \left[\dfrac{1}{8}u^4 + \dfrac{1}{12}u^6\right]_0^{\sqrt{3}}$

$= \dfrac{9}{8} + \dfrac{27}{12} = \dfrac{27}{8}$

🅔 The process of integration by substitution is often done well but some candidates are puzzled here when the process does not immediately complete the change of variable. In part (a) the use of $\sin^2 x = 1 - \cos^2 x$ is usually managed but part (b) causes more problems. Accurate solutions in both parts indicate work of grade-A standard.

Question 8

(i) Let $\dfrac{7 - 2x - 29x^2}{(1 - x)(2 + x)(1 - 3x)} \equiv \dfrac{A}{1 - x} + \dfrac{B}{2 + x} + \dfrac{C}{1 - 3x}$

Hence $7 - 2x - 29x^2 \equiv A(2 + x)(1 - 3x) + B(1 - x)(1 - 3x) + C(1 - x)(2 + x)$
Putting $x = 1$: $-24 = A \times -6$ giving $A = 4$
Putting $x = -2$: $-105 = B \times 21$ giving $B = -5$
Putting $x = \frac{1}{3}$: $\frac{28}{9} = C \times \frac{14}{9}$ giving $C = 2$

Therefore $f(x) \equiv \frac{4}{1 - x} - \frac{5}{2 + x} + \frac{2}{1 - 3x}$

(ii) $f(x) = 4(1 - x)^{-1} - \frac{5}{2}\left(1 + \frac{1}{2}x\right)^{-1} + 2(1 - 3x)^{-1}$

$= 4(1 + x + x^2) - \frac{5}{2}\left(1 - \frac{1}{2}x + \frac{1}{4}x^2\right) + 2(1 + 3x + 9x^2)$

$= \frac{7}{2} + \frac{45}{4}x + \frac{171}{8}x^2$

(iii) The separate expansions are valid for $-1 < x < 1$, $-2 < x < 2$ and $-\frac{1}{3} < x < \frac{1}{3}$ respectively. So the expansion for $f(x)$ is valid for $-\frac{1}{3} < x < \frac{1}{3}$.

Part (i) is generally completed successfully, although some candidates make avoidable careless slips. The expansion of $\frac{5}{2 + x}$ in part (ii) causes problems; candidates often rewrite the expression as $5(2 + x)^{-1}$ and then adjust incorrectly to $10\left(1 + \frac{1}{2}x\right)^{-1}$. Part (iii) is testing. Candidates who realise that it is the smallest of the ranges of values corresponding to the three separate expansions that is appropriate for $f(x)$ are displaying mathematical awareness of grade-A standard.

Question 9

(i) Separating the variables: $\int \frac{dA}{A - 4} = \int -0.2t\,dt$, giving $\ln(A - 4) = -0.1t^2 + c$

When $t = 0$, $A = 60$, giving $\ln 56 = c$

Hence $\ln(A - 4) = -0.1t^2 + \ln 56$ or $\ln \frac{A - 4}{56} = -0.1t^2$

Therefore $\frac{A - 4}{56} = e^{-0.1t^2}$, leading to $A = 4 + 56e^{-0.1t^2}$

(ii) Substituting $A = 10$: $6 = 56e^{-0.1t^2}$ giving $-0.1t^2 = \ln \frac{6}{56}$ and hence $t = 4.73$

(iii) The area infested approaches $4\,m^2$.

Most candidates can separate the variables, although some make the subsequent integration more awkward by obtaining $\int \frac{dA}{0.2(4 - A)} = \int t\,dt$. Most obtain an integral involving natural logarithms but rearranging to make A the subject of a formula is not an easy process for many. Some marks are available for an incorrect expression for A. Not many candidates can assess what the model indicates about the long term.

Answers to exercises

Exercise 1

(1) $\dfrac{x+3}{2x+1}$

(2) $\dfrac{5}{6}(y-1)$

(3) $t+5$

Exercise 2
(1) Quotient $= 4x + 10$, remainder $= 11x - 27$
(2) Quotient $= x^2 + 7x - 1$, remainder $= -9x - 4$
(3) $2x^4 - 13x^3 + 10x^2 + 16x - 3$

Exercise 3

(1) $\dfrac{6}{2x-1} + \dfrac{2}{x+3}$

(2) $-\dfrac{3}{x-2} + \dfrac{9}{(x-2)^2} + \dfrac{3}{x+1}$

(3) $-\dfrac{1}{8y} + \dfrac{37}{40(3y-4)} + \dfrac{3}{20(y+2)}$

Exercise 4
(1) $1 - 8x + 40x^2 - 160x^3$

(2) $\dfrac{1}{4} - \dfrac{3}{128}x + \dfrac{27}{8192}x^2;\ -\dfrac{16}{3} < x < \dfrac{16}{3}$

(3) $1 + \dfrac{1}{2}x + \dfrac{1}{4}x^2;\ 16$

Exercise 5

(1) $\dfrac{t-3}{t}$

(2) $1 + 6x + 16x^2$

(3) $\dfrac{2}{3(x-1)} - \dfrac{2}{3(x+2)} - \dfrac{2}{(x+2)^2}$

(4) $2 + \dfrac{1}{4}x - \dfrac{1}{64}x^2$

(5) (i) $\dfrac{8}{2+x} + \dfrac{16}{1-2x};\ 20 + 30x + 65x^2$ (ii) -2

Exercise 6

(1) $-\dfrac{8}{3}$

(2) $6xe^{2x} + 12x^2e^{2x} + 4x^3e^{2x}$

(3) $2x - 7y + 19 = 0$

(4) $-5, -1$

Exercise 7

(1) (i) $2x \tan 5x + 5x^2 \sec^2 5x$

    (ii) $\dfrac{1 + 2\sin 2x}{x - \cos 2x}$

    (iii) $\dfrac{2e^{2x}(2\sin x - \cos x)}{\sin^2 x}$

(3) $y = -\dfrac{1}{2}x + \dfrac{1}{12}\pi + 2\sqrt{3}$

Exercise 8

(1) $xy^2 = 6 + y^2$

(2) $(0, 5), (0, 3), (0, 1), (-15, 0)$

Exercise 9

(1) $y = 7x + 72$

(3) $t = 3; \left(\dfrac{3}{4}e^3, -8\right)$

Exercise 10

(1) $6x + 5y + 15 = 0$

(2) $\dfrac{8 - 3x^2}{3\cos 3y}$

Exercise 11

(1) (i) $12\cos(2x + \tfrac{1}{3}\pi)$ **(ii)** $3\tan^2 x \sec^2 x$ **(iii)** $2\sec 4x \tan 4x$

(2) $\ln 2$

(3) $4x^2 - 9y^2 + 18y = 8$

(4) (i) $(2, -7), (2, 5)$ **(ii)** $(-1, -1), (5, -1)$

(5) $8\sqrt{3}$

Exercise 12

(1) (i) $2x^3 - 3x^2 + c$ **(ii)** $\tfrac{1}{4}\ln|4x + 7| + c$ **(iii)** $\tfrac{2}{3}e^{3x} + 4e^{-2x} + c$

(2) $18\tfrac{1}{4}$

(3) $2e^5 - 2$

(4) 30 cubic units

Exercise 13

(1) $4\ln\tfrac{5}{3}$

(2) $-\dfrac{2}{x} - \ln|x| + \ln|2x - 1| + c$

(3) $4\ln|x + 1| + 2\ln|x - 1| - \ln|x + 2| + c$

A2 Mathematics

Exercise 14
(1) 2
(2) 15

Exercise 15
(1) $\frac{1}{2}x + \frac{1}{12}\sin 6x + c$
(2) $6\tan \frac{1}{2}x - 3x + c$
(3) $\frac{1}{6}\pi - \frac{1}{4}\sqrt{3}$

Exercise 16
(1) $\frac{1}{7}\ln|\sin 7x| + c$
(2) $\frac{5}{2}\ln|e^{2x} + 4x + 3| + c$
(3) $\ln \dfrac{2\sqrt{2}}{\sqrt{2} + 1}$ or $\ln(4 - 2\sqrt{2})$

Exercise 17
(1) $2x^4 \ln x - \frac{1}{2}x^4 + c$
(2) 16
(3) $-4xe^{-\frac{1}{2}x} - 8e^{-\frac{1}{2}x} + c$
(4) $4e^2$

Exercise 18
(1) $(e^{2x} + 3)^{\frac{1}{2}} + c$
(2) $\frac{45}{112}$
(3) $\frac{2}{35}$

Exercise 19
(1) Resolve the integrand into partial fractions
(2) Integration by parts with $u = \ln x$ and $\dfrac{dv}{dx} = x^5$
(3) Use identity to express $\frac{1}{2}\sin^2 4x$ as $\frac{1}{4} - \frac{1}{4}\cos 8x$
(4) Integration by parts with $u = 2x$ and $\dfrac{dv}{dx} = \sin 5x$
(5) Integrand is of the form $\dfrac{\frac{1}{2}f'(x)}{f(x)}$
(6) Factorise the denominator and resolve $\dfrac{x^2 + 1}{x^2(x - 4)}$ into partial fractions
(7) Integration by parts with $u = \ln 3x$ and $\dfrac{dv}{dx} = 1$
(8) Answer can be written down immediately
(9) Answer can be written down immediately
(10) Expand $(\cos x + \sin x)^2$ and use identities $\cos^2 x + \sin^2 x \equiv 1$ and $2\cos x \sin x \equiv \sin 2x$

(11) Integrand is of the form $\dfrac{f'(x)}{f(x)}$

(12) Integration by parts carried out twice

(13) Resolve the integrand into partial fractions

(14) Answer can be written down immediately

(15) Rewrite the integrand as $\dfrac{1}{\sec x} + \sec^2 x$ and use the fact that $\dfrac{1}{\sec x} \equiv \cos x$

Exercise 20
(1) $4x \tan x + 4 \ln|\cos x| + c$

(2) $\dfrac{15}{8} + \dfrac{9}{4} \ln 2$

(3) $2e^{2x}(2x^2 - 2x + 1) + c$

(4) $\sqrt{3}$

(5) $\dfrac{2}{1 + \sin 2x}; 1$

(7) $\dfrac{3}{4}\pi + \dfrac{9}{8}\sqrt{3}$

Exercise 21
(1) $y^3 = 6e^{2x} + c$

(2) $y = A\left(\dfrac{1}{2}x - 3\right)^2 - 2$

(3) $\ln|\sin\theta| = 4te^{2t} - 2e^{2t} + c$

Exercise 22
(1) $\sqrt{2y + 3} = 2x^2 - 13$

(2) $P = 5t^2 + 6$

Exercise 23
(1) $\dfrac{dM}{dt} = -kM$

(2) $\dfrac{dV}{dt} = -kV^{\frac{1}{3}}$

Exercise 24
(1) $N = 75e^{0.470t}$; 786

(2) (i) $A = \left(\dfrac{1}{15}t^{\frac{3}{2}} + 1\right)^2$

   (ii) 48.5 m^2

   (iii) 36.7

(3) $y^3 = 36\ln|x + 2| - 18\ln|x + 1| + c$

(4) (ii) $\theta = 138e^{kt} + 22$

   (iii) 46.4 minutes

Exercise 25
(1) $5\sqrt{2}$

(2) $a = -3, b = 3\dfrac{2}{3}$

A2 Mathematics

Exercise 26

(1) $\mathbf{r} = \begin{pmatrix} 1 \\ -7 \\ 2 \end{pmatrix} + \lambda \begin{pmatrix} 1 \\ 11 \\ -2 \end{pmatrix}$ or $\mathbf{r} = \begin{pmatrix} 2 \\ 4 \\ 0 \end{pmatrix} + \lambda \begin{pmatrix} 1 \\ 11 \\ -2 \end{pmatrix}$ or $\mathbf{r} = \begin{pmatrix} 1 \\ -7 \\ 2 \end{pmatrix} + \lambda \begin{pmatrix} -1 \\ -11 \\ 2 \end{pmatrix}$ or $\mathbf{r} = \begin{pmatrix} 2 \\ 4 \\ 0 \end{pmatrix} + \lambda \begin{pmatrix} -1 \\ -11 \\ 2 \end{pmatrix}$

(3) (−4, 1, 2)

Exercise 27
(1) 8
(2) 67.8°
(3) 64.7°

Exercise 28
(1) (5, −6, −1)
(2) 76.1°
(3) $m = 4$, $p = 34$; (−3, 9, 26)
(4) (9, −1, 1); 63.0°

Exercise 29
(1) (0.59514, 7.28068)
(2)

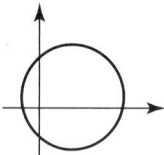

(3) $\frac{4}{3}\sqrt{3}$

Exercise 30
(1) $-k \sin kx$
(2) $\frac{1}{2} - \frac{1}{2} \cos 2\theta$
(3) $\int \pi y^2 \, dx$
(4) $\ln \frac{p}{q}$
(5) $(x-a)^2 + (y-b)^2 = r^2$
(6) $\frac{1}{k} e^{kx} + c$
(7) $1 + \tan^2 \theta = \sec^2 \theta$
(8) $\sqrt{x^2 + y^2 + z^2}$
(9) $-\frac{1}{k} \cos kx + c$
(10) $\frac{1}{2} + \frac{1}{2} \cos 2\theta$
(11) $1 + \cot^2 \theta = \text{cosec}^2 \theta$
(12) $b^2 - 4ac > 0$
(13) $\frac{1}{a} \ln |ax + b| + c$

(14) $-\dfrac{1}{a}(ax+b)^{-1} + c$

(15) $2\sin\theta\cos\theta$

(16) $k\cos kx$

(17) $\dfrac{3}{4}\pi$

(18) $\dfrac{1}{k}\sin kx + c$

(19) Product rule

(20) $\dfrac{-b \pm \sqrt{b^2 - 4ac}}{2a}$

Exercise 31

(1) $\dfrac{1}{8}\pi$ cubic units

(2) **(i)** $x^2 + y^2 - 2x + 8y - 3 = 0$
 (ii) centre $(1, -4)$, radius $= 2\sqrt{5}$

(3) $\dfrac{58}{3}$

(4) $1 + \dfrac{1}{4}(y+5)^2 = \dfrac{1}{9}(x-4)^2$

(5) $1 - 12x + 72x^2$; 7.8103